Praise for the Teach Yourself VISUALLY Series

I just had to let you and your company know how great I think your books are. I just purchased my third Visual book (my first two are dog-eared now!) and, once again, your product has surpassed my expectations. The expertise, thought, and effort that go into each book are obvious, and I sincerely appreciate your efforts. Keep up the wonderful work!

—Tracey Moore (Memphis, TN)

I have several books from the Visual series and have always found them to be valuable resources.

—Stephen P. Miller (Ballston Spa, NY)

Thank you for the wonderful books you produce. It wasn't until I was an adult that I discovered how I learn—visually. Although a few publishers out there claim to present the material visually, nothing compares to Visual books. I love the simple layout. Everything is easy to follow. And I understand the material! You really know the way I think and learn. Thanks so much!

—Stacey Han (Avondale, AZ)

Like a lot of other people, I understand things best when I see them visually. Your books really make learning easy and life more fun.

—John T. Frey (Cadillac, MI)

I am an avid fan of your Visual books. If I need to learn anything, I just buy one of your books and learn the topic in no time. Wonders! I have even trained my friends to give me Visual books as gifts.

—Illona Bergstrom (Aventura, FL)

I write to extend my thanks and appreciation for your books. They are clear, easy to follow, and straight to the point. Keep up the good work! I bought several of your books and they are just right! No regrets! I will always buy your books because they are the best.

—Seward Kollie (Dakar, Senegal)

Credits

Acquisitions Editor
Pam Mourouzis

Project Editor
Suzanne Snyder

Copy Editor
Kelly Henthorne

Technical Editor
Tom Page

Editorial Manager
Christina Stambaugh

Publisher
Cindy Kitchel

Vice President and Executive Publisher
Kathy Nebenhaus

Interior Design
Kathie Rickard
Elizabeth Brooks

Cover Design
José Almaguer

Dedication

This book is dedicated to the memory of my mother-in-law, Margaret Smolinski, who never met a flower, piece of music, or person, and rarely a garage sale, she didn't like. Her great grandchildren knew her as Grandma Birdie. Whenever I picked up one of her telephone calls I was greeted by "Hello, David, this is your mother-in-law, whether you like it or not!" She will never be forgotten.

It is also dedicated to Jakob and Alex Cherry, Francesco, Sebastian, and Gino Nicholas Bubba, Hailee Foster, Rocio, Myles, Kira, Reese, and Zoe Herzog, Marcelo, all their parents, and, of course, Uncle Ian.

About the Author

David Alan Herzog is the author of numerous books in mathematics and other subjects, and over 100 educational software programs in various fields. He taught math education at Fairleigh Dickinson University and William Paterson College, was mathematics coordinator for New Jersey's Rockaway Township public schools, and taught mathematics in New York City's public schools. He was an associate of the federally funded Madison Project of Webster College and Syracuse University, and the Central Iowa Low Achievers Mathematics Project.

Acknowledgments

It is only appropriate that I acknowledge Wiley acquisitions editor Pam Mourouzis who broached the idea of this volume to me. It seemed at the time that it would be a lot of fun. It turned out to be a lot of . . . work! Until you've tried it, you have no idea of how laborious it is to keep shifting the COLOR of the letters that you are typing, but now I have, and I do. Still, I trust that the result will have been worth that labor. I must also acknowledge the labor of project editor Suzanne Snyder, for whom this is the fifth project of mine in a row that she has good-naturedly helped to put together. This was no mean feat, since for the first time I exceeded my scheduled allotment of time, thereby putting her into an even more than usual frenetic time crunch. Then there's Christina Stambaugh who was in charge of managing such things as getting the interior design created, and Cynthia Kitchel, my publisher, who went out of her way to accommodate me and went so far as to phone me to say that she likes my work. Last, but not least, I need to thank my wife, Karen, who promised that she'd make my life miserable if I did not.

Table of Contents

chapter 1 *The Basics*

chapter 2 *Signed Numbers*

chapter 3 — Fractions, Decimals, and Percents

chapter 4 *Variables, Terms, and Simple Equations*

chapter 5 *Axioms, Ratios, Proportions, and Sets*

Monomials, Binomials, and Systems of Equations

Polynomials and Factoring

Cartesian Coordinates

chapter 9 Inequalities and Absolute Value

chapter 10 Algebraic Fractions

Roots and Radicals

Quadratic Equations

chapter 13 Algebraic Word Problems

chapter 1

The Basics

Algebra is a very logical way to solve problems—both theoretically and practically. You need to know a number of things. You already know arithmetic of whole numbers. You will review the various properties of numbers, as well as using powers and exponents, fractions, decimals and percents, and square and cube roots. I am a firm believer in a picture being worth a thousand words, so I illustrate anything it's possible to illustrate and make extensive use of color. Each chapter concludes with practice exercises to help you to reinforce your skills. The solutions to these exercises can be found at the end of the chapter, just before the Glossary.

The most practical chapter in this book is undoubtedly the last one, because it brings together all the skills covered in earlier chapters and helps you put them to work solving practical word problems. You may not believe me right now, but algebra is all about solving word problems—some of them very practical problems, such as, "How much tax will I pay on a purchase? How big a discount is 35% off of something already selling for 25% off? And how good is the gas mileage my motor vehicle is getting?"

Groups of Numbers

Before beginning to review or to learn algebra, it is important to feel comfortable with some pre-algebra concepts including the various groups or realms of numbers with which you will work and the commonly used mathematical symbols and conventions. The first ones you look at are the various families of numbers.

COUNTING NUMBERS

Counting numbers are also known as *natural numbers*. You use these numbers to count things. In the time of cave people, counting numbers probably consisted of one, two, and many and were used to describe how many woolly mammoths had just gone by. As people settled into lives based on agriculture, the need for more specifics arose, so more modern folks got the familiar 1, 2, 3, 4, 5, and so on, as shown here.

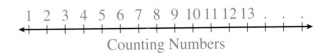

WHOLE NUMBERS

Notice the two arrowheads at either end of the Counting Numbers line. They indicate that the line continues for an infinite length in both directions—infinite meaning "without end." Adding a 0 to the left of the 1 takes you into the realm of **whole numbers**. Whole numbers are not very different from counting numbers. As the number line shows, Whole Numbers = Counting Numbers + 0.

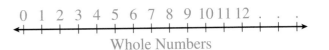

INTEGERS

Integers are the next realm. **Integers** are numbers
into which 1 can be divided without any remainder
being left. Whether or not you've ever done algebra
before, you have come across positive and negative
values. You've seen them on a thermometer. The
temperature inside your house should always be
positive, and around 68 degrees, but in the winter,
the reported outside temperature in Alaska (or
outside your home) often falls below 0. Those are
also known as *negative temperatures*. Similarly, the
stock market may close up or down on any given
day. If it makes gains for the day, it's positive; if the
market loses value, it's negative.

Notice that every integer has a sign attached to it
except for the integer 0. Zero is neither positive nor
negative, but separates one group from the other.
Additionally, notice that +4 is exactly the same
distance from 0 on its right as –4 is on its left.

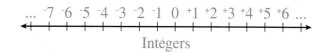

Integers

RATIONAL NUMBERS

You might think from the name that rational numbers are numbers that make sense—and of course, they do—but that's not why they're called rational numbers. A **ratio**, from which the name comes, is a comparison between two quantities.

Say that you have two automobiles, but your neighbor has three. Then the ratio of your cars to your neighbor's is $\frac{2}{3}$. The ratio of your neighbor's cars to yours, on the other hand, is $\frac{3}{2}$.

If you're thinking "I've seen those before, but I called them 'fractions,'" you're absolutely right. A ratio is a form of fraction. Any number that can be expressed as a fraction is a **rational number**. That includes terminating decimal fractions, such as 0.25; percent fractions, such as 25% (each of which could be expressed as $\frac{1}{4}$); repeating decimals; and any integer (since 5 can be expressed as a fraction: $\frac{5}{1}$). So the realm of rational numbers includes all realms that have come before—including $0 \left(\frac{0}{8} = 0 \right)$. You will examine ratios and rational numbers much more fully later.

Just for the record, an example of a repeating decimal is the fraction $\frac{1}{6}$. When it is expressed as a decimal, it is 0.16666666..., which never stops repeating. For simplification, you put a bar over the number that repeats and simplify it to $0.1\overline{6}$. I have not attempted to include a number line for rational numbers for a very good reason: An infinite quantity of rational numbers exists between any two integers.

IRRATIONAL NUMBERS

Don't believe that irrational numbers are so angry that they are unable to think rationally. **Irrational numbers** come about when a number cannot be expressed as either an integer or a rational number. Two examples of irrational numbers are $\sqrt{2}$ (the square root of 2) and π (pi). The value of π has been worked out by computer to be 3 followed by 256 or more decimal places with no repeat occurring. As was the case with rational numbers, an infinite quantity of irrational numbers can fit between any two integers.

PRIME NUMBERS

A **prime number** is a number that has *exactly* two factors, itself and 1. (**Factors** is a name given to numbers that are multiplied together.) The first prime number is 2, with factors of 1 and 2. Next is 3, with factors of 1 and 3. Can you think of any other numbers you could multiply together to make 2 or 3? What are the next two prime numbers? Four is not prime, since it has three factors: 1, 2, and 4. In fact, 2 is the only even prime number. (We examine that further in a paragraph or so.) The next two primes after 3 are 5 and 7. Then come 11, 13, 17, 19, 23, 29, 31, 37, . . .

Why is it important to know prime numbers? The best answer I can give is it's a timesaver—especially when simplifying fractions. If you recognize 19 as a factor of a fraction, and you realize that 19 is a prime number, then you won't waste time trying to simplify it further.

EVEN NUMBERS

Every second number is an **even** one: 2, 4, 6, 8, 10, . . . Counting by 2s is something you probably could do even before you could add or subtract. Any number with a 2, 4, 6, 8, or 0 in its ones place is an even number. Note also that all even numbers contain 2 as a factor. That's why no even number except the first can be prime. Also note that 0 acts like an even number. The realm of even numbers is infinite.

ODD NUMBERS

Every second number is an **odd** one. Huh?! Didn't I just say that every second number is an even one? Strange as it may seem, that's not a contradiction. The only difference is the starting point. To name even numbers you start with 2; you start naming odd numbers with 1. If the ones place of a number contains a 1, 3, 5, 7, or 9, it's an odd number. The main feature of odd numbers is that dividing them by 2 will *not* result in a whole number.

Ways of Showing Things in Algebra

SHOWING MULTIPLICATION

Prior to algebra, multiplication has always been shown with a $\times$ sign. In algebra there are many different ways to show multiplication, but $\times$ is not one of them.

One of the more popular ways to indicate multiplication is with a **multiplication dot**, as shown at right.

$$2 \cdot 3 = 2 \times 3$$

There are also three different ways (see right) to show multiplication of two numbers using parentheses.

$$(3)(5) = 15$$
$$3(5) = 15$$
$$(3)5 = 15$$

We haven't yet studied variables, but I'm sure it's no secret to you that algebra often uses letters to stand for numbers.

Let's use the variables a, b, and c to stand for three different numbers. We could multiply these variable numbers by constant numbers in any of the ways shown above, or we could use the following notations:

$3a$ means 3 times a.

$4b$ means 4 times b.

$5c$ means 5 times c.

We could also use ac to indicate a times c, ab to indicate a times b, or bc to indicate b times c.

Do you think we could show 2 times 3 by writing 23? Sorry, but that configuration is already known as the quantity twenty-three.

COMMON MATH SYMBOLS

Certain mathematical symbols are used throughout this book. It is important that you are familiar with and feel comfortable with them.

Common Math Symbols	
Symbol	**Meaning**
=	is equal to
≠	is not equal to
; or ≈	is approximately equal to
>	is greater than
≥	is greater than or equal to
<	is less than
≤	is less than or equal to

The last four symbols above are the ones with which people often have trouble. I know of some people who remember them as representing alligator jaws, always ready to snap on the larger of the two quantities. In the case of the two types of greater-than symbols, the number on the left is greater; for the two less-than symbols, the greater quantity is on the right, so reading from left to right, the first number is less than the second. A number that is greater than or less than another is a group of numbers that excludes both named quantities from the group. A number that is greater than or equal to a number includes the lower number in the group. One that is less than or equal to a number includes the upper number in the group.

GROUPING SYMBOLS

When writing algebraic expressions and or mathematical sentences (which you'll learn more about in Chapter 4), it is often essential to group certain numbers together, either for the sake of clarity, or in order to specify the sequence in which those numbers should be operated upon. Certain symbols and rules govern these groupings.

Ways of Showing Things in Algebra *(continued)*

PARENTHESES

The most commonly used grouping symbols are **parentheses** (). The equation at the right is a true statement, as you'll see if you solve what's in parentheses on both sides of the = sign:

$$(4 + 3) = (11 - 4)$$
$$7 = 7$$

BRACKETS

When it is necessary to group numbers in parentheses with another group of numbers in parentheses, you use **brackets** [ ]. When you use grouping symbols, you must first clear them by combining the innermost numbers, then working your way outward. In the sentence (below, right) the innermost numbers are $(3 + 4)$ and $(2 + 2)$.

Is the equation a true statement? By combining the numbers in the parentheses you get: $3[7 - 4] = 9$. Next, you perform the indicated subtraction, to get $3[3] = 9$. Brackets perform the same function as parentheses, so $3[3]$ means the same thing as $3(3)$, which you should recognize as indicating multiplication. That leaves your fully interpreted mathematical sentence to read $9 = 9$. Thus, $3[(3 + 4) - (2 + 2)] = 9$ has proven to be a true statement.

$$3[(3 + 4) - (2 + 2)] = 9$$
$$3[7 - 4] = 9$$
$$3[3] = 9$$
$$9 = 9$$

BRACES

Sometimes, although rarely, it is necessary to group numbers together that already are in bracketed parentheses. Then we use grouping symbols known as **braces** { }. Consider the equation at the right.

$$6\{3 + [(6 - 4) + (7 - 5)]\} = 42$$
$$6\{3 + [2 + 2]\} = 42$$
$$6\{3 + 4\} = 42$$
$$6\{7\} = 42$$
$$42 = 42$$

Let's find out whether this mathematical sentence is true or not. As always, start inside the parentheses first, then move to the brackets, and finally the braces.

First solve $(6 - 4)$ and $(7 - 5)$. Next, add the 2s. Then combine the $3 + 4$. Finally, multiply.

Remember, the order of use is always parentheses, then brackets, then braces: $\{[( )]\}$. Sometimes larger parentheses are used as well:

$$(\{[( )]\})$$

Properties and Elements

There are certain properties that apply to addition and multiplication but do not apply to subtraction or division. You probably came across them in elementary school math, but they are unlikely to be fresh in your memory, so let's review them here.

COMMUTATIVE PROPERTY

The **commutative property** applies to addition and multiplication. It deals with order. Formally stated, it says that when two numbers are added together or when two numbers are multiplied together, the order in which they are added or multiplied does not affect the result. Here's a shorter statement:

The commutative property for addition: $(a + b) = (b + a) = a + b$

The commutative property for multiplication: $(a \cdot b) = (b \cdot a) = ab$

At this point in your algebra career, you might better understand it as follows:

The commutative property for addition: $(5 + 3) = (3 + 5) = 8$

The commutative property for multiplication: $(5 \cdot 3) = (3 \cdot 5) = 15$

ASSOCIATIVE PROPERTY

The **associative property** also applies to addition and multiplication. It deals with grouping. In order to understand it, you must first understand that all arithmetic operations are **binary**. That means that you can operate on only two numbers at any one time. This may seem to fly in the face of what you learned about column addition. Remember the columns of four or five numbers, say $3 + 5 + 4 + 9 + 8$ all stacked up over a line segment, as shown here:

$$
\begin{array}{r}
3 \\
5 \\
4 \\
9 \\
8 \\
\hline
\end{array}
$$

Well, the fact of the matter is, you never added more than two of those numbers together at a time. If you tried adding that column right now, you'd probably go 3 + 5 = 8; 8 + 4 = 12; 12 + 9 = 21; 21 + 8 = 29. No contradiction there! You add numbers two at a time, no matter how many there are to add, which brings you back to the associative property for both addition and multiplication. They say that when three (or more) numbers are to be added together or when three or more numbers are multiplied together, the order in which they are grouped for addition or multiplication does not affect the result.

The associative property for addition: $(a + b) + c = a + (b + c) = a + b + c$
The associative property for multiplication: $(a \cdot b) \cdot c = a \cdot (b \cdot c) = abc$

That may also be shown as:

The associative property for addition: $(5 + 3) + 2 = 5 + (3 + 2) = 10$
The associative property for multiplication: $(5 \cdot 3) \cdot 2 = 5 \cdot (3 \cdot 2) = 30$

Remember, the commutative property deals with order, the associative property deals with grouping.

IDENTITY ELEMENTS

The **identity element** is the number that may be combined with another number without changing its value. For addition, the identity element is 0.

$$0 + 3 = 3$$
$$n + 0 = n$$

What do you suppose the identity element is for multiplication? That's right, it's 1. It's demonstrated here:

$$5 \cdot 1 = 5$$
$$1 \cdot y = y$$

For subtraction, the identity element is once again 0.

$$7 - 0 = 7$$
$$n - 0 = n$$

For division, the undoing of multiplication, the identity element is once again 1. Remember, the fraction line symbolizes division.

$$\frac{5}{1} = 5$$
$$\frac{g}{1} = g$$

DISTRIBUTIVE PROPERTY

The **distributive property** is used to remove grouping symbols (usually parentheses) by sharing a multiplier outside the grouped numbers with those on the inside. It always involves numbers that are added together or subtracted within the grouping symbols. For that reason, it is often referred to as the **distributive property of multiplication over addition**.

In the first example at the right, notice that the 3 was distributed.

$$3(5) + 3(2) = 15 + 6$$

The second example involves subtraction.

$$4(6) - 4(3) = 24 - 12$$

Exponents and Powers

An exponent is a number written small to the right and high off the line next to another number. In 2^3, 3 is the exponent, and 2 is the base. An exponent expresses the power to which a number is to be raised (or lowered). (I discuss the lowering power of certain exponents in a later chapter.)

What They Do

An exponent after a number tells the number of times the base number is to be multiplied by itself (see examples at right).

$$4^2 = 4 \cdot 4$$
$$2^3 = 2 \cdot 2 \cdot 2$$
$$6^4 = 6 \cdot 6 \cdot 6 \cdot 6$$

Many would argue that an exponent and a power are the same thing, but they really are not. An exponent is a symbol for the power (how many of the base number are multiplied together). The first example would be read *four to the second power*, or *four squared* (more about that in a moment). The second depicts *two to the third power*, or *two cubed* (more about that in a moment, as well). The third shows *six to the fourth power*.

The exponents 2 and 3 have special names, based upon a plane and a solid geometric figure. Because the square's area is found by multiplying one side by itself, raising something to the second power is called squaring. Since the volume of a cube is found by multiplying one side by itself and by itself again, raising something to the third power is called cubing. These terms are in common use and chances are that you've heard them before. A list of the first 12 perfect squares (squares of whole numbers) is shown at the right.

$1^2 = 1$	$5^2 = 25$	$9^2 = 81$
$2^2 = 4$	$6^2 = 36$	$10^2 = 100$
$3^2 = 9$	$7^2 = 49$	$11^2 = 121$
$4^2 = 16$	$8^2 = 64$	$12^2 = 144$

To cube a number, multiply it by itself and then do it again. A list of the first 12 **perfect cubes** (cubes of whole numbers) is shown at the right.

$1^3 = 1$	$5^3 = 125$	$9^3 = 729$
$2^3 = 8$	$6^3 = 216$	$10^3 = 1000$
$3^3 = 27$	$7^3 = 343$	$11^3 = 1331$
$4^3 = 64$	$8^3 = 512$	$12^3 = 1728$

Comparing squares and cubes to whole numbers, you should note that squares get big very quickly, but not nearly as quickly as cubes.

Any number raised to a power of 1 is that number itself: $3^1 = 3$; $13^1 = 13$.

Any number raised to a power of 0 is equal to 1: $5^0 = 1$; $18^0 = 1$. This is explained further in a following section.

Operations Using Exponents and Powers

It is possible to perform all four arithmetic operations on numbers raised to powers, but some very specific rules apply. You can examine them as they apply to each of the operations, starting with multiplication.

MULTIPLYING

To multiply numbers with the same base, all you have to do is keep the base the same and add the exponents:

$3^2 \cdot 3^2 = 3^{2+2} = 3^4$ $3^4 = 81$ Check it: $\rightarrow 3^2 \cdot 3^2 = 9 \cdot 9 = 81$

$5^3 \cdot 5^3 = 5^{3+3} = 5^6$ $5^6 = 15,625$ Check it: $\rightarrow 5^3 \cdot 5^3 = 125 \cdot 125 = 15,625$

$2^4 \cdot 2^3 = 2^{4+3} = 2^7$ $2^7 = 128$ Check it: $\rightarrow 2^4 \cdot 2^3 = 16 \cdot 8 = 128$

Suppose that the bases are different, such as 2^3 and 4^2? When the bases are different, each number must be expanded. First find 2^3 which is $2 \cdot 2 \cdot 2 = 8$; 4^2 is $4 \cdot 4 = 16$. Then multiply the results to get $8 \cdot 16 = 128$.

2^3 and $4^2 =$ ___
$2 \cdot 2 \cdot 2 = 8$
$4 \cdot 4 = 16$
$8 \cdot 16 = 128$

You'll find practice examples at the end of the chapter on page 18.

DIVIDING

To divide numbers with the same base, all you have to do is keep the base the same and subtract the exponents. Does that sound like the exact opposite of multiplication? You bet:

$$3^3 \div 3^2 = 3^{3-2} = 3^1 \qquad 3^1 = 3 \qquad \text{Check it:} \rightarrow 3^3 \div 3^2 = 27 \div 9 \quad = 3$$
$$5^4 \div 5^2 = 5^{4-2} = 5^2 \qquad 5^2 = 25 \qquad \text{Check it:} \rightarrow 5^4 \div 5^2 = 625 \div 25 = 25$$
$$2^6 \div 2^3 = 2^{6-3} = 2^3 \qquad 2^3 = 8 \qquad \text{Check it:} \rightarrow 2^6 \div 2^3 = 64 \div 8 \quad = 8$$

Here's a bonus explanation. It's the one promised two sections back. This is why any number raised to the 0 power = 1. Let n stand for any number, or if you'd prefer, you can substitute any number you like for n. The results will be the same. In the first line to the right, you see $n^3 \div n^3$. But that is a number divided by itself. Any number divided by itself = 1, so n^0, which you see in the second line, equals 1. Pretty cool, don't you think?

$$n^3 \div n^3 = n^{3-3}$$
$$n^{3-3} = n^0$$

As with multiplication, if the bases are different, you have to expand each value before combining. For example, to find $4^2 \div 2^3$, you first find the values of each, 16 and 8; then you divide: $16 \div 8 = 2$.

ADDING AND SUBTRACTING

To add or subtract numbers with exponents, whether the bases are the same or not, each expression must be evaluated (expanded) and then the addition or subtraction found. Answers are at the end of the chapter.

Try These

❶ $2^3 + 3^2 =$ _____

❷ $2^4 + 5^3 =$ _____

❸ $3^3 + 4^2 =$ _____

❹ $6^2 - 3^2 =$ _____

❺ $9^2 - 3^3 =$ _____

❻ $12^2 - 4^3 =$ _____

RAISING TO ANOTHER EXPONENT

When a number with an exponent is raised to another exponent, simply keep the base the same as it was, but multiply the exponents.

$$\left(3^4\right)^3 = 3^{12} \quad \left(2^3\right)^3 = 2^9 \quad \left(4^2\right)^2 = 4^4 \quad \left(7^3\right)^8 = 7^{24} \quad \left(5^6\right)^3 = 5^{18}$$

Square Roots and Cube Roots

Operating with square roots and cube roots in algebra is explored later in this book. For now, I am concerned with defining them, identifying them, and simplifying them.

IDENTIFYING SQUARE ROOTS AND CUBE ROOTS

The **square root** of a number is the number that when multiplied by itself gives that number. The square root or radical sign looks like this: $\sqrt{\ }$. You read $\sqrt{25}$ as "the square root of 25." A list of the first 12 **perfect (whole number) square roots** is given at right.

$$\sqrt{0} = 0 \qquad \sqrt{1} = 1 \qquad \sqrt{4} = 2$$
$$\sqrt{9} = 3 \qquad \sqrt{16} = 4 \qquad \sqrt{25} = 5$$
$$\sqrt{36} = 6 \qquad \sqrt{49} = 7 \qquad \sqrt{64} = 8$$
$$\sqrt{81} = 9 \qquad \sqrt{100} = 10 \qquad \sqrt{121} = 11$$

Square roots may also be identified using a fractional exponent $\frac{1}{2}$. For example, an alternative to writing the square root of 144 is shown at right.

$$\sqrt{144} = 144^{\frac{1}{2}} = 12$$

The **cube root** of a number is the number that when multiplied by itself twice gives that number. The cube root bracket looks like this: $\sqrt[3]{\ }$. You read $\sqrt[3]{27}$ as "the cube root of 27." A list of the first 12 **perfect (whole number) cube roots** is given at right.

$$\sqrt[3]{0} = 0 \qquad \sqrt[3]{1} = 1 \qquad \sqrt[3]{8} = 2$$
$$\sqrt[3]{27} = 3 \qquad \sqrt[3]{64} = 4 \qquad \sqrt[3]{125} = 5$$
$$\sqrt[3]{216} = 6 \qquad \sqrt[3]{343} = 7 \qquad \sqrt[3]{512} = 8$$
$$\sqrt[3]{729} = 9 \qquad \sqrt[3]{1000} = 10 \qquad \sqrt[3]{1331} = 11$$

Cube roots may also be identified using a fractional exponent $\frac{1}{3}$. For example, an alternative to writing the cube root of 1728 is shown at right.

$$\sqrt[3]{1728} = 1728^{\frac{1}{3}} = 12$$

Square Roots and Cube Roots (continued)

SIMPLIFYING SQUARE ROOTS

Not all numbers have perfect square roots; in fact, most of them do not. Some square roots, however, can be simplified. Take, for example, $\sqrt{40}$. It is possible to do some factoring beneath the radical sign.

$$\sqrt{40} = \sqrt{4 \cdot 10} = 2\sqrt{10}$$

Note that you can factor a perfect square (4) out of the 40. Then you can remove the square 4 from under the radical sign, writing its square root in front of the sign, so $\sqrt{40}$ becomes 2 times $\sqrt{10}$.

Do you see any way to simplify $\sqrt{45}$? Look for a perfect square that is a factor of 45. 9 is such a number. The square root of 9 is 3, so remove the 9 from under the radical sign and write a 3 in front of it. The simplest form of $\sqrt{45}$ is $3\sqrt{5}$.

$$\sqrt{45} = \sqrt{9 \cdot 5} = 3\sqrt{5}$$

APPROXIMATING SQUARE ROOTS

For the majority of numbers that are not perfect squares, it is sometimes necessary to *approximate* a square root. Take, for example, $\sqrt{57}$. $49 < 57 < 64$. That means 57 is between 49 and 64, or greater than 49 and less than 64. $\sqrt{57}$ should fall between 7 and 8, the square roots of 49 and 64, respectively. 57 is about halfway between 49 and 64, so try 7.5. $7.5^2 = 56.25$; try a little higher: $7.6^2 = 57.76$. The number you're looking for is about halfway between both of those results, so try halfway between. $7.55^2 = 57.0025$. Now that's pretty darned good.

Generally, the square roots of nonperfect squares can be found using a table or a calculator that has a square root function. You may want to commit these two to memory:

$$\sqrt{2} \approx 1.414 \qquad \sqrt{3} \approx 1.732$$

Zero, FOO, and Divisibility

MULTIPLYING AND DIVIDING BY ZERO

$0 \cdot$ **anything** $= 0$. Maybe you already knew that, but I can't emphasize it too much. If you have 5 apples 0 times, how many apples do you have? The answer is 0. How much is $5,000 \cdot 0$? Anything times 0 is 0. Remember the commutative property for multiplication from earlier in this chapter. Division by zero does not exist. Anything divided by 0 is **undefined**. Since a fraction is, among other things, a division, **the denominator of a fraction can never equal 0**.

FUNDAMENTAL ORDER OF OPERATIONS

If parentheses, addition, subtraction, exponents, multiplication, and so on are all contained in a single problem, the order in which they are done matters. A mnemonic device to help remember that order is P.E.M.D.A.S., or Please Excuse My Dear Aunt Sally. The letters really stand for Parentheses Exponents Multiplication Division Addition Subtraction and represent the sequence in which the operations are to be performed—with certain provisos, as indicated below:

1. Parentheses
2. Exponents or square roots
3. Multiplication ⎫
4. Division ⎬ Whichever comes first from left to right.
5. Addition ⎫
6. Subtraction ⎬ Whichever comes first from left to right.

DIVISIBILITY TESTS

You can save time when factoring (or dividing) numbers by following these rules.

A number is divisible by	If
2	Its ones digit is divisible by 2.
3	The sum of its digits is divisible by 3.
4	The number formed by its two right-hand digits is divisible by 4.
5	Its ones digit is a 0 or a 5.
6	It is divisible by 2 and by 3 (use the rules above).
7	Sorry, there's no shortcut here.
8	The number formed by its three rightmost digits is divisible by 8.
9	The sum of its digits is divisible by 9.

Chapter Practice

Practice Questions

GROUPS OF NUMBERS

1 What is the meaning of the arrowheads on the end of the number lines representing each of the realms of numbers?

2 How do integers differ from rational numbers?

3 Name two irrational numbers.

COMMON MATH SYMBOLS

4 Write an is-greater-than symbol followed by an is-greater-than-or-equal-to symbol.

5 Name the conventional algebraic grouping symbols from innermost to outermost.

6 Represent 5 + the product of 7 and 4 in mathematical symbols.

7 Represent 5 is about equal to 4.9 in mathematical symbols.

COMMUTATIVE AND ASSOCIATIVE PROPERTIES

8 Demonstrate the commutative property for multiplication.

9 Demonstrate the associative property for addition.

IDENTITY ELEMENTS AND DISTRIBUTIVE PROPERTY

10 Name the identity element for subtraction.

11 Name the identity element for division.

12 Demonstrate the distributive property by filling in the blank: $a(b + c) =$ _____

13 Demonstrate the distributive property by filling in the blank: $5(7 - 3) =$ _____

POWERS AND EXPONENTS

14 What is the meaning of 2^4?

15 What is the meaning of 4^3?

16 Represent 9 squared.

17 Represent 7 cubed.

OPERATIONS USING POWERS AND EXPONENTS

Express each answer in exponential form.

18 $3^2 \cdot 3^5$

19 $4^3 \cdot 4^6$

20 $9^4 \div 9^3$

21 $12^8 \div 12^5$

22 $6^3 + 6^4$

Evaluate each of the following expressions.

23 $3^2 \cdot 5^4$

24 $6^3 + 6^4$

25 $5^3 + 2^5$

26 $6^3 - 3^4$

27 $5^2 - 4^3$

SQUARE ROOTS AND CUBE ROOTS

28 Name the first four perfect squares and their square roots.

29 Name the first four perfect cubes and their cube roots.

30 Express $\sqrt{50}$ in simplest terms using the radical sign.

31 Express $\sqrt{128}$ in simplest terms using the radical sign.

32 Approximate the square root of 42.

FUNDAMENTAL ORDER OF OPERATIONS AND DIVISIBILITY TESTS

33 What is the mnemonic device for remembering the order of operations?

34 Express in simplest form: $3 + [10(9 - 2^2)]$

35 Express in simplest form: $20 - 2 \cdot 6 + 12^2 + (9 - 1) \cdot 4$

36 When is a number divisible by 6?

37 Is 17,438,156,241 divisible by 9?

Chapter Practice Answers

1 They indicate that the values continue infinitely in either direction.

2 Integers are whole numbers both positive and negative, and rational numbers include non-whole numbers, such as $\frac{1}{2}$ or $\frac{3}{4}$.

3 $\sqrt{2}$, $\sqrt{3}$, and π, to name a few.

4 $>$, $\geq$

5 parentheses, brackets, braces; ( ), [ ], { }

6 $5 + (7 \cdot 4)$

7 $5 \approx 4.9$

8 $(a \cdot b) = (b \cdot a)$ or (number 1 $\cdot$ number 2) = (number 2 $\cdot$ number 1)

9 $(a + b) + c = a + (b + c)$ or (number 1 + number 2) + number 3 = number 1 + (number 2 + number 3)

10 0

11 1

12 $a(b + c) = ab + ac$

13 $5(7 - 3) = 35 - 15 = 20$

14 $2 \cdot 2 \cdot 2 \cdot 2 = 16$

15 $4 \cdot 4 \cdot 4 = 64$

16 9^2

17 7^3

18 3^7

19 4^9

20 9

㉑ 12^3

㉒ $6^3 + 6^4$ — I hope you didn't fall for that one.

㉓ 5625

㉔ 1512

㉕ 157

㉖ 135

㉗ −39

㉘ $\sqrt{0} = 0$ $\quad \sqrt{1} = 1$ $\quad \sqrt{4} = 2$ $\quad \sqrt{9} = 3$

㉙ $\sqrt[3]{0} = 0$ $\quad \sqrt[3]{1} = 1$ $\quad \sqrt[3]{8} = 2$ $\quad \sqrt[3]{27} = 3$

㉚ $5\sqrt{2}$

㉛ $8\sqrt{2}$

㉜ 6.5

㉝ Please Excuse My Dear Aunt Sally. *PemDAS Exponents multiplication Addition subtraction*

㉞ First clear the parentheses by squaring the 2 and subtracting the resulting 4 from 9 to get 3 + [10(5)]; next multiply 10 times 5 to get 3 + 50; finally, add to get 53.

㉟ First clear parentheses by subtracting 1 from 9: $\qquad$ $20 - 2 \cdot 6 + 12^2 + 8 \cdot 4$

Next remove the exponent by squaring the 12: $\qquad$ $20 - 2 \cdot 6 + 144 + 8 \cdot 4$

Next do the multiplications from left to right: $\qquad$ $20 - 12 + 144 + 32$

Finally, take 12 from 20, and add everything else: $\qquad$ $8 + 144 + 32 = 184$

㊱ When it is divisible by 3 and by 2.

㊲ $1 + 7 + 4 + 3 + 8 + 1 + 5 + 6 + 2 + 4 + 1 = 42$; 42 is not divisible by 9. No.

Chapter Practice
(continued)

ANSWERS FROM P. 14

1. 17

2. 141

3. 43

4. 27

5. 54

6. 80

chapter 2

Signed Numbers

Signed numbers are a basic tool of algebra. In fact, every number you'll ever deal with in algebra is a signed number, whether or not you can see that sign. In this chapter, I help you to build an understanding of signed numbers, which are not as esoteric as they might appear at first blush. Unless you live somewhere like Florida, you've likely come across them at one time or another in the weather forecast on a particularly wintry day. You've also possibly experienced them on your bank statement when you have a negative balance from inadvertently spending more than was in your account. With those two cheerful thoughts in mind, here's hoping that you're reading this with a positive temperature on the thermometer and a positive bank balance.

Introducing Signed Numbers

It is not uncommon for people to have difficulty comprehending the concept of signed numbers and their universality. For that reason, the Madison Project—a federally funded math-teaching improvement project with which I had the honor of being associated—set out to find many everyday experiences in which to couch this basic building block of algebra. It is with that purpose in mind that the devices of balloons and sandbags, letter carrier stories, and the apartment house (which immediately follows) came to exist. The hope was and is that being able to relate this totally theoretical fact with something that can be related to in everyday life will facilitate your ability to understand the concept.

An Apartment Analogy

A large apartment building was built on the side of a hill. It looked remarkably like the building shown on the following page. The main floor was the entry level of the building, and to go from floor to floor you rode in an elevator. Of course, to get to the floor you wanted, you needed to press the correct button in the elevator.

In the town where this building was located, an ordinance said that for a floor to be numbered with a positive number (1, 2, 3, and so on), that floor had to be completely above ground level. As you can see on the following page, some of the floors were completely above ground level, and others were not.

The entrance level could have been called "main floor" or "lobby," but since the elevator buttons were so tiny, the manufacturer chose to put a 0 on that button. All the numbers above ground level were labeled by which level above the ground they were and were prefixed by a positive sign (+), just for emphasis. The floors below entry level were labeled according to how many levels below the entry level they were and were prefixed by a negative sign (−).

How great is the distance between the +3 level and the 0 level? How about the −3 level and the 0 level? I hope you said "3" in response to each of those questions. If you're not sure, use the number line on page 30, place a fingernail on 0 and count the number of spaces you cross to get to +3. Then do the same thing with 0 and −3. How great is the distance between the +5 level and the −5 level? Count it out just to make sure. Did you get 10? Are you positive? That last one was kind of a trick question, but if a number doesn't have a negative sign in front of it, it's considered positive. Remember that. It'll come in handy.

What's the difference between the 9th floor and the −8th floor? Here's where the last sentence in the previous paragraph proves to be true. The 9th floor is 9 levels above 0; the −8th floor is 8 levels below 0. That's a total difference of 17 floors.

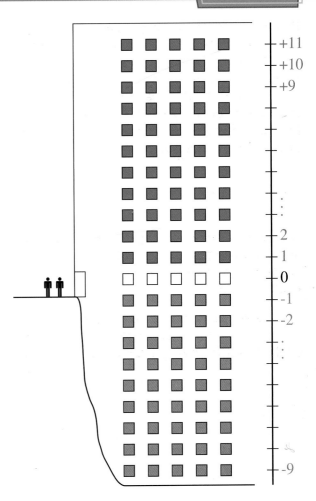

Adding Balloons and Sandbags

A hermit named Frank lived in a small cabin on the edge of a cliff. The ground around Frank's cabin was quite sandy, so it just sat on that sandy ground without having any kind of foundation.

Frank was quite happy living there in peace and quiet. Then, suddenly, he was awakened one morning by the clatter and roar of heavy construction machinery digging in the valley below. After several hours of thinking and grumbling, Frank went to the store, returning several hours later carrying a fat package under one arm and wheeling a cylinder of helium in front of him. After some minutes of fussing, Frank attached two helium-filled balloons to the roof of his house. One was a +8 balloon, and the other was a +6 balloon, as you can see here.

The balloons had the effect of lifting the house right off of the cliff. How high above the cliff do you think the house went? You should be able to draw the following conclusion: Frank added a +8 balloon; that's + (+8), which is so named because it elevated the house 8 feet. Then he added a +6 balloon; that's + (+6), which elevated the house 6 more feet.

+ (+8) + (+6) = +14, or a total elevation of 14 feet.

The additional 14 feet of height between Frank's house and the construction machinery deadened the sound just enough to enable Frank to keep hanging around the neighborhood (pun intended).

After two or three days of hanging around 14 feet above the cliff, Frank realized that he was running out of groceries. He couldn't just jump out and go shopping, but thankfully, he had prepared for such an eventuality on the day he had brought home the helium cylinder. Buried in the package under his arm was a long, flexible plastic tube and a battery-powered suction pump. Being resourceful, he weighted down one end of the tube and unrolled it out a window until it reached the ground. Then he connected the other end of the tube to the suction pump and sucked up a good quantity of sand (you'll recall that his house originally sat on quite sandy ground). He used the sand to fill a couple of sandbags, which he hung. One was a –9 sandbag and the other was a –6 sandbag.

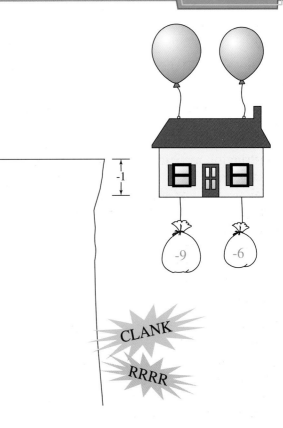

The sandbags had the effect of lowering the house below the cliff. How far below the edge of the cliff do you think the house ended up? You should be able to draw the following conclusion:

Frank added a –9 sandbag; that's + (–9), which is so named because it lowered the house 9 feet. Then he added a –6 sandbag; that's + (–6), which lowered the house 6 more feet.

Therefore: + (–9) + (–6) = –15

But since the house was 14 feet above the cliff to begin with, 14 + (–15) gives it a total elevation of –1 feet, or 1 foot below the edge of the cliff. That allows Frank to put out a board and then walk onto the cliff-top to go grocery shopping.

Subtracting Balloons and Sandbags

One day, Frank the hermit was sitting in his cabin floating serenely above the cliff with many balloons and sandbags attached to his cabin, and he realized that he was getting low on flashlight batteries. He was, in fact, so low on battery power that he couldn't operate the pump he usually used to collect sand in order to lower his house. What was he to do? He pondered for several hours, until suddenly he came up with a plan. Getting out a pair of shears with a very long reach, he stretched out of his window and cut a balloon off of the house. It turned out to have been a +7 balloon. What do you think happened as a result?

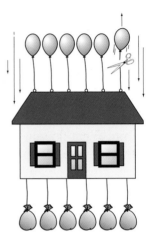

The house dropped exactly 7 feet.

"My word," Frank exclaimed. "Cutting off a balloon has the same effect as adding a sandbag." Next, he stretched out of his window and cut another balloon off of the house. It was a +10 balloon. What do you think happened as a result?

Hopefully, you realized that the house dropped exactly 10 feet more. Let's represent the events of the last few paragraphs mathematically:

Cutting off two balloons of specified values:
$$-(+7) - (+10) = -17$$

That says that removing a +7 balloon and removing a +10 balloon causes the house to drop 17 feet.

Another day, Frank had just returned from shopping and there were lots of balloons and sandbags attached to the house. He was about to inflate a couple of balloons to get away from the din in the valley below when he discovered that he was out of helium. This time, he didn't even hesitate; he got out his shears and cut off a −9 sandbag from beneath the cabin. What do you think was the result of that? That's right—the house was lifted +9 feet.

Not content with being so low, Frank cut off another −9 sandbag from beneath the cabin. Where did the cabin end up?

Cutting off two sandbags of the specified amounts:
$$-(-9) - (-9) = +18$$

If you can think about the next section in terms of adding or removing balloons or sandbags and the attendant results, you will be going a long way toward understanding the basics of algebra.

Understanding Number Lines

You looked briefly at number lines in Chapter 1 when examining the various groups of numbers. In algebra, every number has a sign associated with it, except for 0.

As you can see on this integer number line, 0 separates the positive integers from the negative ones. A negative number is always preceded by a negative sign (–). If a number has no sign in front of it, it is understood to be positive (+). Both signs are considered separate and apart from the operational signs minus and plus (– and +), which indicate subtraction and addition, respectively.

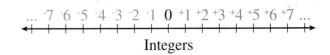

Integers

Among the conventions you need to know about any number line is that if you see arrowheads at either end, they indicate that the sequence represented on the line goes on forever in the direction of that arrowhead. Numbers increase in value from left to right and decrease in value from right to left. That means any number is greater than the one to the left of it and smaller than the one to its right, *regardless of their signs*.

Notice that it is also possible to have number lines containing fractions and/or mixed numbers. The other essential convention on a number line is that the intervals between any two numbers must be the same, whether counting by halves (as in this figure to the right), by ones (as in the above figure), or by twos, fives, or tens.

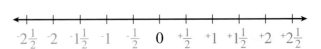

Here's where it might help you to think back to Frank the hermit and his balloons and sandbags. To add any two positive numbers together, simply add their values together and append a positive sign (or not).

For example: $+2 + +5 = +7$, or just plain 7.

To add any two negative numbers together, simply add their values together and append a negative sign.

For example: $-3 + -6 = -9$.

Notice that the negative sign *must* be there.

Adding a positive to a negative gets a little tricky. That's because the result depends on which quantity has the greater absolute value. The **absolute value** is the distance of a number from 0. Look at the top figure on the previous page, and you'll see that $+3$ and -3 have the same absolute values. Each is 3 units from 0. In the bottom figure on the previous page, $+2\frac{1}{2}$ has the same absolute value as $-2\frac{1}{2}$. The statement $|-5| = 5$ is read "the absolute value of negative 5 equals 5."

Now consider this: $+7 + -3 =$ ___.

Because the signs are different, the two quantities are going to be subtracted. What? That's an addition sign; how can you subtract? Go back to the hermit for a moment. Suppose he adds a $+7$ balloon and a -3 sandbag to the house at the same time. What would the net result be? You have an upward force of $+7$ acting against a downward force of -3. That's a net upward force of $+4$. Where did that +4 come from? From subtraction, of course. But how can you subtract when it says . . . ?

Now cut that out!

If it helps you, you can think of rewriting $+7 + -3 =$ ___ as $+7 - +3 =$ ___; just exchange the two signs. Now here's a final thought on the subject: The answer takes the sign of the *addend* (the number being added) with the larger absolute value. Because $+7$ has a greater absolute value than -3, the answer is $+4$.

Try these:

$$\begin{array}{cccc} +5 & -7 & +4 & -8 \\ \underline{-9} & \underline{+4} & \underline{-3} & \underline{+5} \\ -4 & -3 & +1 & -3 \end{array}$$

Answers: $-4, -3, +1, -3$

Subtracting Signed Numbers

You've subtracted signed numbers before; you just never realized you were doing it.

Look at $27 - 14 = 13$. You just didn't know that they were positives you were subtracting. Look at the same subtractions with the signs in place:

$$+27 - +14 = +13$$

To subtract one positive signed number from another positive signed number, switch the inner signs: $+27 + -14 = \_\_$. Minus a positive becomes plus a negative. You may be asking how you can do that, and I'm going to have to ask you to take my word for it *for the moment*. It will become evident when you deal with signed number multiplications.

Look again at the new, rewritten subtraction:

It's now an addition, and you already studied how to handle that. You subtract the absolute values, and the answer takes the sign of the number with the larger absolute value: $+27 + -14 = +13$.

$$+27 + -14 = \_\_$$

How would you handle this subtraction?

$$-12 - +8 = \_\_$$

It starts off just like the last one, with an exchange of the inner two signs:

$$-12 + -8 = \_\_$$

Now what? It *should* look familiar. It's an addition of two negatives (two sandbags, if you will). When adding two negatives, you add the absolute values and give the sum a negative sign. Piece of cake!

$$-12 + -8 = -20$$

> **TIP**
>
> Did your mother, father, or teacher ever point out to you that a double negative is a positive? It certainly is in the English language! What does the next sentence mean? "I am not not going to the fair." It means "I am going to the fair." The two *not*'s cancel each other out. The same is true with signed numbers, so watch out for the double negative.

Look at this subtraction:

$$+8 - -6 = \underline{\quad}$$

From the tip, you can conclude that the double negative becomes a positive, therefore:

You might argue that it became a double positive, so I'll concede and make it $+8 + 6 = \underline{\quad}$. Remember, there doesn't have to be a positive sign to make it a positive number.

$$+8 - -6 = \underline{\quad} \quad \text{becomes} \quad +8 + +6 = \underline{\quad}.$$

Next, combine the 6 and 8 to get 14, so:

$$+8 - -6 = 14$$

Now consider the last possible unique case:

$$-9 - -8 = \underline{\quad}$$

Once again, a double negative must be dealt with, and you should know how to do that:

$$-9 - -8 = \underline{\quad} \quad \text{becomes} \quad -9 + 8 = \underline{\quad}$$

Because the signs are different now, you solve the equation by subtracting the absolute values and keeping the sign of the addend with the greater absolute value. The sum is -1.

Minus Sign before Parentheses

The only rule that makes subtraction of signed numbers different from addition is the Change the Signs rule. After the sign of the number being subtracted has been changed, the subtraction becomes an addition of signed numbers and follows all the various rules for addition.

Sometimes you'll find a minus sign in front of parentheses, such as $-(-6+5-3+3-4)$. The solution is to change all the signs within the parentheses and change the minus to a plus; then add:

$$-(-6+5-3+3-4) \quad \text{becomes} \quad +(+6-5+3-3+4)$$

The parentheses are now meaningless, so:

Notice that not all plus signs or addition signs were written. Sometimes the signs speak for themselves and indicate what's to be done. Sometimes the positive signs are omitted altogether and are replaced by addition signs. That may take some getting used to.

$$+6-5+3-3+4 = +1+0+4$$
$$+1+0+4 = +5, \text{ or } 5$$

TIP

It is very common to think about numbers getting bigger the farther they get from zero, and this is very true of positive numbers. That is, 12 is bigger than 3, 47 is bigger than 12, and so forth. But, all things considered, it might not be quite so obvious that the opposite is true when dealing with negative numbers. That is to say, numbers get smaller as they get farther away from zero. Think about it. –3 is smaller than –1, –12 is smaller than –3, and –64 is smaller than –12. Conversely, the closer a positive number gets to zero, the smaller it is, yet the closer a negative number gets to zero, the larger it is. It can be kind of mind boggling until you put it into perspective.

When these first were devised by the Madison Project they were known as "Postman Stories." Isn't it wonderful how far political correctness has brought us!

Once upon a time, there was a rather eccentric letter carrier who never paid much attention to the names on the envelopes she delivered. In a day or two she'd come back and say, "I'm sorry, Mr. Smith, but I need to take back the mail I gave you the other day; those envelopes were for Mr. Brown and Ms. Ortega."

Mr. Smith didn't pay attention to the names on the checks and bills that the letter carrier brought. A check is a draft on a bank that you may consider income. A bill is a statement asking you for money, such as one from the electric company.

When Mr. Smith receives checks in the mail, he considers himself to be richer by the face amounts of those checks. When he receives bills, he considers himself to be poorer by the amounts of those bills, since once paid, that's money he no longer has in his account. Let's see how that works.

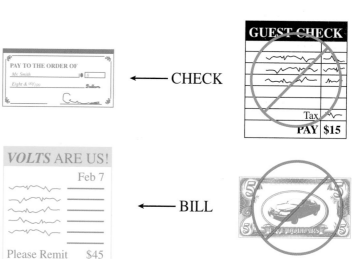

← CHECK

← BILL

One day, the letter carrier brought 4 checks for $7 each. Mathematically, that's represented this way:

+4	×	+7	= ___
↑		↑	↑
She		checks	for a
brought		for	total of
4		$7 each	how much?

Clearly this is a multiplication. In order to figure out what the sign of the product will be, you need to ask yourself: "As a result of the letter carrier's visit, will Mr. Smith consider himself to be richer or poorer?" The letter carrier brought checks, so Mr. Smith believes he is better off by the amount of money those checks are worth. That means that the product will be positive. Now multiply the two numbers and get a result of +28. That means he believes himself to be better off by $28.

....

On another occasion, the letter carrier brought Mr. Smith 3 bills for $16 each. We represent that mathematically like this:

+3	×	−16	= ___
↑		↑	↑
She		bills	for a
brought		for	total of
3		$16 each	how much?

To figure out what the sign of the product will be, you again need to ask yourself, "As a result of this letter carrier's visit, will Mr. Smith consider himself to be richer or poorer?" The letter carrier brought bills, so Mr. Smith believes that he is poorer by the amount of money those bills will cost. That means that the product will be negative. Now multiply the two numbers and get a result of −48. That means he believes himself to be poorer by $48 as a result of this visit.

The letter carrier had to return to Mr. Smith's house to tell him that those 5 checks for $8 each had not really been intended for him, and she had to take them back. Fortunately, Mr. Smith hadn't tried to cash them—fortunately for the letter carrier, who needed to correct her mistake, and for Mr. Smith, who might have found himself arrested for fraud. The equation to the right represents mathematically the letter carrier's visit that morning:

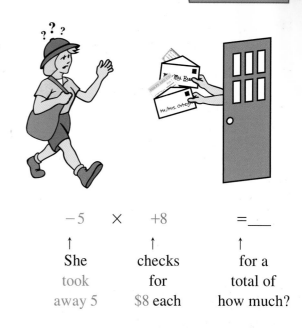

As a result of this visit, will Mr. Smith consider himself to be richer or poorer, and by how much? That shouldn't be too hard to figure out. He's having checks taken away from him. That's money he thought he had, so he's going to consider himself poorer. That means a negative sign. How much poorer? Multiply 5 · 8. The answer is –40, or poorer by $40.

$$-5 \quad \times \quad +8 \quad = \underline{\quad}$$

$$\uparrow \qquad\qquad \uparrow \qquad\qquad \uparrow$$

She took away 5 · checks for $8 each · for a total of how much?

Are you ready for the pièce de resistance?

On another day, the letter carrier comes and tells Mr. Smith that she's there to reclaim the 6 bills for $12 each that she had delivered the day before. They weren't meant for Mr. Smith after all. This transaction is represented mathematically here:

$$-6 \quad \times \quad -12 \quad = \underline{\quad}$$

$$\uparrow \qquad\quad \uparrow \qquad\qquad \uparrow$$

She took away 6 · bills for $12 each · for a net change of how much?

How does Mr. Smith consider himself to have been affected by the letter carrier's visit? Does he believe himself to be richer, poorer, or exactly the same as before she came? Bills were taken away. That meant money he thought he would have to pay can now go back into his pocket. Is that a positive or a negative experience? Having more money always struck me as a positive thing, so the answer is +72.

You should be able to draw some conclusions from these letter carrier stories that will serve you well in understanding the multiplication and division of signed numbers. Keep this model in mind as you go through the next section.

Multiplying Signed Numbers

Only two conditions for multiplying signed numbers are possible, even though four were covered in the letter carrier stories on the preceding pages. If you relate the conditions to the letter carrier stories, you should be able to understand them.

First, consider two positives. Multiplying two positive numbers is the same as the letter carrier bringing checks. The result is positive.

Now consider two negatives. That is the same as the letter carrier taking away bills. Again, the result is positive. Can you draw a conclusion from those two instances? You betcha! When multiplying two numbers, if the signs are the same, the product is positive. Remember also, that you can never multiply more than two numbers at a time. If three numbers are being multiplied, you must multiply two of them together first and then multiply the product by the third number. (This is the associative property, which was covered in Chapter 1.)

Next, consider the possible situations involving two different signs. If you have a positive times a negative, that's the same as the letter carrier bringing bills—a decidedly negative situation. Now try the reverse, a negative times a positive. That is analogous to the letter carrier's taking back checks; that's not something to look forward to, either, and so is another negative. What conclusion can you draw from these two examples? When multiplying two numbers, if the signs are different, the product is negative.

TRY THESE

1 $+5 \cdot +4 = \underline{+20}$

2 $-8 \cdot -3 = \underline{+24}$

3 $+7 \cdot -5 = \underline{-35}$

4 $-6 \cdot +8 = \underline{-48}$

5 $-3 \cdot +7 = \underline{-21}$

6 $+5 \cdot -9 = \underline{-45}$

7 $-4 \cdot -6 = \underline{+24}$

8 $+9 \cdot +8 = \underline{72}$

9 $+7 \cdot -11 = \underline{-77}$

10 $-6 \cdot -9 = \underline{+54}$

11 $+8 \cdot +7 = \underline{+48}$

12 $-10 \cdot +12 = \underline{-120}$

The color coding should have helped out. You'll have to pay more attention during the Chapter Practice. For these, if the colors of both factors are the same, the products are positive; if they are different, they're negative.

ANSWERS

1. +20
2. +24
3. −35
4. −48
5. −21
6. −45

7. +24
8. +72
9. −77
10. +54
11. +56
12. −120

Dividing Signed Numbers

You're going to find this section very anticlimactic, but then again, your brain could probably use a break. Make sure that you understand the rules for multiplying signed numbers, because the rules for dividing them are exactly the same. The one exception is that you divide instead of multiply.

TRY THESE

1 $+20 \div +4 =$ ___

2 $-18 \div -3 =$ ___

3 $+35 \div -5 =$ ___

4 $-64 \div +8 =$ ___

5 $-56 \div +7 =$ ___

6 $+81 \div -9 =$ ___

7 $-42 \div -6 =$ ___

8 $+72 \div +8 =$ ___

ANSWERS

1 $+5$

2 $+6$

3 -7

4 -8

5 -8

6 -9

7 $+7$

8 $+9$

$$\left\{ \begin{array}{l} + \times + \\ - \times - \end{array} \right\} \longrightarrow + \qquad \left\{ \begin{array}{l} + \div + \\ - \div - \end{array} \right\} \longrightarrow +$$

$$\left\{ \begin{array}{l} + \times - \\ - \times + \end{array} \right\} \longrightarrow - \qquad \left\{ \begin{array}{l} + \div - \\ - \div + \end{array} \right\} \longrightarrow -$$

Practice Questions

Perform the indicated operations and write your answers in the blanks provided. Pay careful attention to the signs.

1. $-9 + -3 = $ ____
2. $+17 + -5 = $ ____
3. $-16 + +8 = $ ____
4. $-13 + +7 = $ ____
5. $+15 + -9 = $ ____
6. $-14 + -6 = $ ____
7. $+9 + +8 = $ ____
8. $+14 + -11 = $ ____
9. $-6 + -9 = $ ____
10. $+18 + +7 = $ ____
11. $-20 + +12 = $ ____
12. $+9 - +5 = $ ____
13. $-7 - -4 = $ ____
14. $+11 - -6 = $ ____
15. $-6 - +8 = $ ____
16. $-13 - +7 = $ ____
17. $+5 - -9 = $ ____
18. $-4 - -16 = $ ____
19. $+11 - -11 = $ ____
20. $-6 - -9 = $ ____
21. $+8 - +7 = $ ____
22. $-10 - +12 = $ ____

㉓ $+7 \cdot +6 =$ _____

㉔ $-9 \cdot -6 =$ _____

㉕ $+8 \cdot -9 =$ _____

㉖ $-7 \cdot +4 =$ _____

㉗ $-5 \cdot +8 =$ _____

㉘ $+6 \cdot -7 =$ _____

㉙ $-5 \cdot -8 =$ _____

㉚ $+4 \cdot -12 =$ _____

㉛ $-8 \cdot -7 =$ _____

㉜ $-9 \cdot +9 =$ _____

㉝ $+10 \div +4 =$ _____

㉞ $-81 \div -3 =$ _____

㉟ $+30 \div -5 =$ _____

㊱ $-72 \div +8 =$ _____

㊲ $-63 \div +7 =$ _____

㊳ $+45 \div -9 =$ _____

㊴ $-15 \div -6 =$ _____

㊵ $+27 \div +3 =$ _____

㊶ $+88 \div -11 =$ _____

㊷ $-81 \div -9 =$ _____

㊸ $+49 \div +7 =$ _____

㊹ $-72 \div +8 =$ _____

Chapter Practice Answers

1 −12

2 +12

3 −8

4 −6

5 +6

6 −20

7 +17

8 +3

9 −15

10 +25

11 −8

12 +4

13 −3

14 +17

15 −14

16 −20

17 +14

18 +12

19 +22

20 +3

21 +1

22 −22

23 +42

24 +54

25 −72

26 −28

27 −40

28 −42

29 +40

30 −48

31 +56

32 −81

33 +2.5

34 +27

35 −6

36 −9

37 −9

38 −5

39 +2.5

40 +9

41 −8

42 +9

43 +7

44 −9

Fractions, Decimals, and Percents

Fractions, decimals, and percents all deal with parts of things and are all, to some extent, interchangeable with one another. As you worked with signed number arithmetic in the last chapter, somewhere in your mind you probably knew that it could not apply only to whole numbers. The time to extend that knowledge is now, as you look at how to handle signed fractions, decimals, and percents and how to exchange one for the other. Don't discount (excuse the pun) the fact that a knowledge of percentage can help you save big money at sales. It's also quite interesting. Finally, scientific notation makes it possible for the scientific community to treat very large and very small quantities in a manageable fashion. You don't have to be a nuclear scientist or a molecular chemist to use them, but it couldn't hurt.

Fractions take many different forms and can represent many different things. When you think of a fraction, you probably picture something like $\frac{1}{2}$, and you define it as a *part of a whole*. In fact, that's only part of the story—and a very small part at that (no pun intended). A fraction may be a ratio, comparing two different things; it may be a division example, as $\frac{4}{2} = 2$; or it may be a whole unto itself, as would be a piece of pie. I know I've never heard anyone ask for $\frac{1}{8}$ of a pie.

The kinds of fractions shown here are called **common fractions** and sport both numerators and denominators. You can't lose sight of the fact that **decimals** and **percents** are also types of fractions. **Decimal fractions** are based on ten tenths equaling one whole. **Percent fractions** are based on 100 representing one whole. You usually don't refer to decimal fractions and percent fractions as such, but rather you call them decimals and percents, respectively; just don't lose sight of the fact that they are types of fractions.

The term *fraction* used by itself in this book refers to common fractions. Fractions can be displayed on a number line, as shown early in Chapter 1. They, as well as integers, may be negative or positive. How to operate with signed fractions of one sort or another is what this chapter is all about.

A FRACTION AS:

Part of a whole

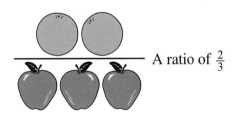

A ratio of $\frac{2}{3}$

$$\frac{9}{3} = 3 \qquad \text{A division}$$

A whole slice of pie; want half of it?

Adding and Subtracting Signed Fractions

Negative fractions can be written in three different ways as shown here. Although all three mean the same thing, the third way is how negative fractions are usually written. Can you see why the first two forms mean the same thing? It's because, as you just saw in the last chapter, a negative divided by a positive and a positive divided by a negative both result in a negative.

$$\frac{-3}{4} = \frac{3}{-4} = -\frac{3}{4}$$

Adding Signed Fractions

When working with signed fractions, the same rules apply that apply to whole numbers, and to nonsigned fractions, for example. Consider the following:

$$-\frac{1}{4} + \frac{1}{3} = \underline{\quad}$$

The first thing that should be apparent is that (from what you know about fractions) you can't add without first finding a common denominator.

$$-\frac{1}{4} + \frac{1}{3} = -\frac{3}{12} + \frac{4}{12} = \underline{\quad}$$

Now that you have both fractions in terms of the common denominator, you apply the rules of adding signed numbers: When the signs are different, you subtract, and the sum takes the sign of the addend with the greater absolute value:

$$-\frac{3}{12} + \frac{4}{12} = \frac{1}{12}$$

From here on, you can avoid writing unnecessary positive signs. Do one more addition:

$$-\frac{1}{2} + \left(-\frac{2}{3}\right) = -\frac{3}{6} + \left(-\frac{4}{6}\right) = -\frac{7}{6} = -1\frac{1}{6}$$

You express halves and thirds in terms of sixths. Then, since the signs are the same, you add the fractions' values together. Finally, you express the sum in the simplest terms.

Subtracting Signed Fractions

Do you remember how to subtract signed whole numbers? Again, the same rules apply. Start with two fractions that already have the same denominator:

$$-\frac{3}{8} - \left(-\frac{1}{8}\right) = \underline{\hspace{1cm}}$$

Do you remember the double negative rule?

$$-\left(-\frac{1}{8}\right) \text{ becomes } +\frac{1}{8}$$

That means the subtraction has become addition:

$$-\frac{3}{8} + \frac{1}{8} = \underline{\hspace{1cm}}$$

Finally, you add and simplify the sum:

$$-\frac{3}{8} + \frac{1}{8} = -\frac{2}{8} = -\frac{1}{4}$$

Of course, if the denominators were different, you would find a common one, convert, and proceed from there. Try one more example:

$$-\frac{5}{6} - \left(+\frac{1}{3}\right) = \underline{\hspace{1cm}}$$

Here, you must rewrite both fractions in terms of the common denominator, which is sixths:

$$-\frac{5}{6} - \left(+\frac{2}{6}\right) = \underline{\hspace{1cm}}$$

Next, you adjust the signs (minus a positive becomes plus a negative), and then, since the signs are the same, you add.

$$-\frac{5}{6} + \left(-\frac{2}{6}\right) = -\frac{7}{6} = -1\frac{1}{6}$$

And there you have it!

Multiplying and Dividing Signed Fractions

Multiplication and division of signed fractions incorporate techniques that you have already studied, both in earlier grades when you learned how to operate with fractional multiplication and in Chapter 2 when you dealt with multiplication and division of signed integers. In this section, you review each of those techniques very briefly and then look at a couple of examples of each technique.

Multiplying Signed Fractions

In case you don't recall, to multiply fractions, you multiply the denominators together and multiply the numerators together, not necessarily in that order, and then simplify the product if necessary, so as to leave it in its simplest form. If possible, you cancel before multiplying so as to make the process easier and avoid having to simplify the product. To multiply signed fractions, you follow both the rules for multiplying fractions and those for multiplying signed integers: Like signs give positive products, and unlike signs yield negative ones. Here are a couple of examples:

$$\frac{3}{4} \times -\frac{1}{6} = \underline{}$$

The first thing to notice is that the signs are different. That means the product is going to be negative. It's also possible to do some canceling:

$$\frac{\cancel{3}^{1}}{4} \times -\frac{1}{\cancel{6}_{2}} = \underline{}$$

Finally, you multiply numerators and denominators:

And that's one down. If you did not understand anything about this example, please study the steps again. You might need to refresh yourself on multiplying signed numbers (Chapter 2) or multiplying fractions (summarized previously).

$$\frac{\cancel{3}^{1}}{4} \times -\frac{1}{\cancel{6}_{2}} = -\frac{1}{8}$$

Here's another example:

$$-\frac{5}{8} \times -\frac{3}{7} = \underline{\quad}$$

Notice right off that the signs are both negative. When the signs are the same, the product is positive. It doesn't look like there's room for canceling, either:

$$-\frac{5}{8} \times -\frac{3}{7} = + \underline{\quad}$$

Multiplying numerators and denominators brings you to the final product. You could have left the positive sign, but it's superfluous.

$$-\frac{5}{8} \times -\frac{3}{7} = +\frac{15}{56} = \frac{15}{56}$$

Dividing Signed Fractions

Division of fractions takes advantage of the fact that division is the reciprocal operation for multiplication. A **reciprocal**, you may recall, is the number by which you multiply another one to get a product of 1. So, the reciprocal of 3 is $\frac{1}{3}$; the reciprocal of $\frac{1}{4}$ is 4; and the reciprocal of $\frac{2}{3}$ is $\frac{3}{2}$.

When dividing one fraction by another, the divisor (second fraction) is replaced by its reciprocal, and the divided-by sign becomes a times sign. You may have learned to perform this logical process by the rote phrase "invert and multiply." From then on, you follow the rules for multiplying fractions and the rules for multiplying signed integers: Like signs give positive products, and unlike signs yield negative ones. Here is an example:

$$-\frac{2}{3} \div \frac{3}{4} = \underline{\quad}$$

① Are you tempted to cancel? Don't do it! Canceling can only be done across a × sign. Step 1 is to change this division to a reciprocal multiplication:

$$-\frac{2}{3} \div \frac{3}{4} = -\frac{2}{3} \times \frac{4}{3} = \underline{\quad}$$

2 Now is when you would cancel if it were possible, which it isn't. You now have a signed fraction multiplication with unlike signs. That means the quotient (the answer in a division) is negative; multiply:

$$-\frac{2}{3} \times \frac{4}{3} = -\frac{8}{9}$$

Look at a second example:

$$-\frac{4}{5} \div \left(-\frac{12}{15}\right) = \_\_\_$$

---

1 First you need to rewrite the division as a reciprocal multiplication:

$$-\frac{4}{5} \times \left(-\frac{15}{12}\right) = \_\_\_$$

---

2 Now that there's a × sign, is it possible to cancel? You betcha! So do it:

$$-\frac{\overset{1}{\cancel{4}}}{\underset{1}{\cancel{5}}} \times \left(-\frac{\overset{3}{\cancel{15}}}{\cancel{12}_{3}}\right) = \_\_\_$$

---

3 Do you see any other canceling that can be done there? Actually, you end up with this:

$$-1 \times -1 = +1 = 1$$

You remembered that two negatives make a positive, right?

Fractions greater than 1 are also known as mixed numbers when expressed with a whole number part followed by a fractional part. Recall that when adding and subtracting mixed numbers, the whole number parts are done together and the fractional parts are done together (added or subtracted). Of course, when the mixed numbers are signed, the normal signed number rules for addition and subtraction must be followed.

Multiplying Signed Fractions Greater Than 1

To multiply mixed numbers, you must first convert each mixed number to a fraction with a numerator larger than the denominator. The conversion is achieved by multiplying the whole number by the fraction's denominator and then adding its numerator to the product. That amount is then put over the fraction's denominator.

$$2\frac{1}{2} \longrightarrow 2\frac{1}{2} = \frac{5}{2}$$

START

$$6\frac{3}{5} \longrightarrow 6\frac{3}{5} = \frac{33}{5}$$

START

Solve the following multiplication:

$$-3\frac{1}{2} \times -2\frac{3}{4} = \underline{\qquad}$$

Convert both mixed numbers:

$$-\frac{7}{2} \times -\frac{11}{4} = \underline{\qquad}$$

Finally, multiply and simplify:

$$-\frac{7}{2} \times -\frac{11}{4} = +\frac{77}{8} = 9\frac{5}{8}$$

Do you get the picture? Notice that the laws governing signed number multiplication were followed.

Multiplying and Dividing Mixed Numbers *(continued)*

Here's one more:

$$4\frac{2}{3} \times -3\frac{3}{5} = \underline{\quad}$$

Convert both mixed numbers:

$$\frac{14}{3} \times -\frac{18}{5} = \underline{\quad}$$

You can cancel:

$$\frac{14}{\cancel{3}_1} \times -\frac{\cancel{18}^6}{5} = \underline{\quad}$$

Finally, multiply and simplify:

$$\frac{14}{1} \times -\frac{6}{5} = -\frac{84}{5} = -16\frac{4}{5}$$

Summarizing, like signed numbers multiply together to give a positive product; unlike signed numbers multiply together to give a negative product. Sounds familiar!

Here's another:

$$-4\frac{3}{8} \div \left(+2\frac{3}{6}\right) = \underline{\quad}$$

First, change to fractions:

$8 \times 4 = 32 + 3 + 35$; and $6 \times 2 = 12 + 3 = 15$, so we get

$$-\frac{35}{8} \div \frac{15}{6} = \underline{\quad}$$

which becomes:

$$-\frac{35}{8} \times \frac{6}{15} = \underline{\quad}$$

Cancel, multiply, and simplify:

$$-\frac{\cancel{35}^7}{\cancel{8}_4} \times \frac{\cancel{6}^3}{\cancel{15}_3} = -\frac{7}{4} \times \frac{1}{1} = -\frac{7}{4} = -1\frac{3}{4}$$

Sometimes fractions contain other fractions in the numerator, other fractions in the denominator, or both. There's an example of each case below.

Let's see how to go about simplifying each of those complex fractions. The best thing to keep in mind as you plan your strategy is *you must not have an addition or subtraction sign within a fraction*. Do whatever you need to do to remove it (remember, I'm talking plus and minus, not to be confused with positive and negative).

a) $\dfrac{-\frac{3}{4}+(-\frac{3}{6})}{5}$ b) $\dfrac{7}{-\frac{5}{8}+(+\frac{1}{3})}$ c) $\dfrac{+2\frac{1}{2}-(-\frac{3}{5})}{-\frac{2}{3}-(-1\frac{5}{8})}$

To get rid of the plus sign in the numerator of this fraction, you're going to have to do some adding:

$$\frac{-\frac{3}{4}+\left(-\frac{3}{6}\right)}{5}=\frac{-\frac{9}{12}+\left(-\frac{6}{12}\right)}{5}=\frac{-\frac{15}{12}}{5}$$

Now you have a negative fraction over a positive integer. But any integer can be expressed as a fraction—in this case $\frac{5}{1}$. What does the following mean?

$$\frac{-\frac{15}{12}}{\frac{5}{1}}$$

Remember, any fraction is also a division example, so $\dfrac{-\frac{15}{12}}{\frac{5}{1}}$ means the same as $-\frac{15}{12}\div\frac{5}{1}$. To solve that, you turn the

$$-\frac{15}{12}\div\frac{5}{1}=-\frac{15}{12}\times\frac{1}{5}=\underline{}$$

division into a multiplication by the reciprocal of the divisor:

You have some canceling opportunities here:

Now that's what I call a solution.

$$-\frac{\overset{3}{\cancel{15}}}{12}\times\frac{1}{\underset{1}{\cancel{5}}}=-\frac{\overset{1}{\cancel{3}}}{\underset{4}{\cancel{12}}}\times\frac{1}{1}=-\frac{1}{4}$$

Simplifying Complex Fractions *(continued)*

Here comes another one. Remember the main rule governing complex fractions (from the top of the previous page)?

$$\frac{7}{-\frac{5}{8}+\left(+\frac{1}{3}\right)}$$

The first thing you need to do is get rid of that addition sign. You do that by finding the LCD for 8ths and 3rds, which is, if I recall correctly, 24ths:

$$\frac{7}{-\frac{5}{8}+\left(+\frac{1}{3}\right)}=\frac{7}{-\frac{15}{24}+\left(+\frac{8}{24}\right)}=\frac{7}{-\frac{7}{24}}$$

Now what are you going to do with +7 over $-\frac{7}{24}$? Keeping in mind that every integer can be expressed as a fraction, how about this?

$$\frac{7}{-\frac{7}{24}}=\frac{\frac{7}{1}}{-\frac{7}{24}}$$

Then you turn that into a division:

$$\frac{\frac{7}{1}}{-\frac{7}{24}}=\frac{7}{1}\div-\frac{7}{24}$$

Last but not least, you solve that division by making it a reciprocal multiplication:

$$\frac{7}{1}\div-\frac{7}{24}=\frac{7}{1}\times-\frac{24}{7}$$

Let's cancel and multiply:

$$\frac{\overset{1}{7}}{1}\times-\frac{24}{\underset{1}{7}}=-24$$

Now that you've seen how to handle the straightforward ones, try dealing with a couple of twists. The main rule from the last page still applies, but its application is a bit trickier with this complex fraction and others of its ilk. For openers, mixed numbers are not helpful as part of a complex fraction.

$$\dfrac{+2\frac{1}{2}-\left(-\frac{3}{5}\right)}{-\frac{2}{3}+\left(-1\frac{5}{8}\right)}$$

The first task is to get rid of the mixed numbers by changing them to fractions with numerators greater than their denominators. Perhaps you've noticed that I've avoided using the term **improper fractions.** If you ever come across it, be aware of the fact that it is an antiquated term for fractions with numerators greater than their denominators. On the other hand, for it to be an improper fraction in my estimate, it would have had to stay out past 1:30 A.M. without having called home. No fractions should ever be left with numerators greater than their denominators (unless instructed to do so), since there is a much more understandable way of writing them (mixed numbers), but they can be very useful as a means to an end. You start by rewriting the mixed numbers above:

$$\dfrac{+2\frac{1}{2}-\left(-\frac{3}{5}\right)}{-\frac{2}{3}+\left(-1\frac{5}{8}\right)}=\dfrac{+\frac{5}{2}-\left(-\frac{3}{5}\right)}{-\frac{2}{3}+\left(-\frac{13}{8}\right)}$$

The new numerator is now an addition (remember the double negative), so there is a reversal of fortune for the second fraction on top:

$$\dfrac{+\frac{5}{2}-\left(-\frac{3}{5}\right)}{-\frac{2}{3}+\left(-\frac{13}{8}\right)}=\dfrac{+\frac{5}{2}+\frac{3}{5}}{-\frac{2}{3}+\left(-\frac{13}{8}\right)}$$

Next, you need to find like denominators for both **terms** of the fraction. (The numerator and the denominator is each a term.) Here it's done step by step:

$$\frac{+\frac{5}{2}+\frac{3}{5}}{-\frac{2}{3}+\left(-\frac{13}{8}\right)} = \frac{+\frac{}{10}+\frac{}{10}}{-\frac{}{24}+\left(-\frac{}{24}\right)} = \frac{+\frac{25}{10}+\frac{6}{10}}{-\frac{16}{24}+\left(-\frac{39}{24}\right)}$$

Now you can perform the addition:

$$\frac{+\frac{25}{10}+\frac{6}{10}}{-\frac{16}{24}+\left(-\frac{39}{24}\right)} = \frac{\frac{31}{10}}{-\frac{55}{24}}$$

Note that the superfluous positive sign was dropped. You have a division here, so perform the appropriate reciprocal multiplication and get it over with. Nice round answer, don't you think?

$$\frac{\frac{31}{10}}{-\frac{55}{24}} = \frac{31}{\cancel{10}_{5}} \times -\frac{\cancel{24}^{12}}{55} = -\frac{372}{275} = -1\frac{97}{275}$$

As you should know, decimals are also a form of fraction, based on dividing 1 by multiples of the number 10. Any decimal can be expressed as a common fraction simply by saying its decimal name and then writing the fraction equivalent.

Here are three examples:

0.3 means three tenths → $\frac{3}{10}$

0.07 means seven hundredths → $\frac{7}{100}$

−0.211 means negative two hundred eleven thousandths → $-\frac{211}{1000}$

Did you think I had forgotten that we're dealing with signed numbers? Not likely! Certain decimals deserve special attention because they simplify to become fractions that we encounter every day. Here are some examples of those:

$$0.125 = \frac{1}{8} \quad \text{not to mention} \quad -0.125 = -\frac{1}{8}$$
$$0.25 = \frac{1}{4} \quad\quad -0.25 = -\frac{1}{4}$$
$$0.375 = \frac{3}{8} \quad\quad -0.375 = -\frac{3}{8}$$
$$0.5 = \frac{1}{2} \quad\quad -0.5 = -\frac{1}{2}$$
$$0.625 = \frac{5}{8} \quad\quad -0.625 = -\frac{5}{8}$$
$$0.75 = \frac{3}{4} \quad\quad -0.75 = -\frac{3}{4}$$
$$0.825 = \frac{7}{8} \quad\quad -0.825 = -\frac{7}{8}$$

The decimals shown here are all terminating decimals, which means that they consist of a certain number of figures and then end. Certain decimals are repeating ones that never end. An example of that is 0.3333... The ellipsis (three dots) following the last written figure indicates that it goes on doing the same thing forever. This is sometimes represented by a bar over the repeating character, as pictured here:

$$0.33\overline{3} \text{ or } -0.33\overline{3}$$

You know what percents look like. They are numbers with those funny-looking "%" squiggles after them. Does the squiggle look like anything you've seen before? How about a fraction ($\frac{0}{0}$)? Well, a percent *is* a fraction, but it's a different kind of fraction. Common fractions and decimal fractions are based on one whole being represented by the number 1. In the case of percent fractions, one whole is represented by the number 100. That is, one whole is 100%. Keep that in the front of your mind as you proceed.

CHANGING FRACTIONS TO DECIMALS

Let's start by changing a fraction to a decimal. To change a fraction to a decimal, do what the fraction says; that is, $\frac{1}{8}$ means 1 divided by 8. Perform that division, and you get 0.125.

Now try changing the fraction $\frac{1}{3}$ to a decimal. Does your result look familiar? It's that repeating decimal with which the last section ended. What do you suppose the decimal equivalent of $-\frac{2}{3}$ is? If you concluded that it is –0.666..., you are correct, but it's rarely written that way. It is much more common to see it as –0.667, or simply –0.67.

CHANGING DECIMALS TO PERCENTS

The method for changing a decimal fraction to a percent fraction is shown in the chart on the following page.

Three things should strike you about this chart. First, the percents are all positive. How is it possible to have a positive percent? If something goes up by a part of itself, that's a positive increase. Yes, there are negative percents also. If something goes down by a part of itself, thet's a negative increase, more commonly known as a decrease.

The second thing that should grab your attention is that 900%. You're undoubtedly aware by now that 100% is the whole thing. But isn't it possible to have 9 whole things? If Reese has a whole pie and Myles has 9 whole pies, then Myles has 900% the pies that Reese has.

The third thing you should have noticed about the chart is that five hundredths (.05) as a percent is 5%. If you think about it, it makes perfectly good sense, but I wanted to call your attention to the fact that the 0 between the decimal point and the 5 has disappeared. All leading zeroes have also disappeared. Just in case you were unaware of the fact before, notice that multiplying by 100 is the same thing as moving the decimal point two places to the right.

Decimal	×100	=%
0.63	×100	63%
0.3	×100	30%
0.05	×100	5%
0	×100	900%

CHANGING PERCENTS TO DECIMALS

The chart at right should come as no surprise to you whatsoever. It is the exact *opposite process* of that used to change a decimal to a percent. This time, note the necessity of placing the 0 between the decimal point and the 9. The zeroes to the left of the decimal points are gratuitous but aid some in reading. Please note, also, that dividing by 100 is the same as moving the decimal point two places to the left.

Percent	÷100	=Decimal
57%	÷100	0.57
40%	÷100	0.40
9%	÷100	0.09
700%	÷100	7

Finding a Percent of a Number

Adding and subtracting percents don't really have much practical use, since each is the same as doing so with ordinary integers. Of much more practical value is finding a percent of another number. Practically speaking, you often need to find the percent discount at a sale (verifying that the amount of money they've discounted an item corresponds to the percent discount the store's advertising), or what percentage of the people you've invited to your next party are likely to actually show up, RSVP or not.

To find a percent of a number, you multiply, but there is a catch. 25% of 16 is 4, but multiplying by a percent can't be done. What was that? Didn't I just say that you find a percent of a number by multiplying? Yes, but I also mentioned a *catch*, and this is it. There is no way to multiply a number by a percent, but all is not as crazy as it might look. You studied, not very far back, changing percents to decimals, and you certainly can multiply by decimals, so . . . 25% = 0.25.

Next we multiply:

Note where the decimal point was placed.

If you didn't remember how to place the decimal point in a multiplication, remember that you count the total number of digits to the right of the decimal in both numbers being multiplied. This time there were two, so there must be the same number of digits to the right of the decimal point in the product. Since they were both zeroes, you can drop them and see that 25% of 16 is 4.

$$
\begin{array}{r}
25\% = 0.25 \\
16 \\
\times .25 \\
\hline
80 \\
320 \\
\hline
4.00
\end{array}
$$

Find a percent of a negative number, say 40% of –80:

$$40\% = 0.40 = 0.4$$

Next we multiply:

Again, note where the decimal point was placed.

$$
\begin{array}{r}
-80 \\
\times\ .4 \\
\hline
-32.0
\end{array}
$$

DISCOUNTS

A discount is a percent removed from a price. If a coat lists for $120 and is being sold at 30% off, then to find the sale price, you would find 30% of $120, which happens to be $36. Next, you would subtract that $36 from the list price of the coat to get a sale price of $120 − 36 = $84.

Find the sale price of a $230 bicycle that is being sold at a 20% discount:

$$20\% = 0.20 = 0.2$$

. . . .

Now we find the discount:

Finally, find the sale price: $230 − 46 = $184.

$$\begin{array}{r} \$230 \\ \underline{\times .2} \\ 46.0 = \$46 \end{array}$$

Understanding Scientific Notation

Scientific notation is used to write a very small or very large number as a number between 1 and 10 multiplied by 10 raised to a positive or negative power, which depends on the direction in which you must move the decimal point. We might even call scientific notation empowering!

VERY LARGE NUMBERS

For a very large number, the decimal point is going to be moved almost all the way to the left. Remember, the first part must be between 1 and 10. 3,400,000 in scientific notation looks like this: 3.4×10^6. You probably know where the 3.4 came from, but what about the 10^6?

Count the number of places that the decimal point was moved to the left, and that's where the 6 came from. You may work out the value of 10^6 if you like, and you'll see that it really works, because when you multiply 3.4 by 10^6 you get $3,400,000$.

Try writing $91,035,000,000$ in scientific notation. Then try writing $801,600,000,000,000$ in scientific notation. Don't look ahead until you've actually tried to do both of them. Cover the rest of the page so you're not tempted to peek.

Here are the solutions. The first one is 9.1035×10^{10}. Note that the decimal point was moved to the left 10 places. In the second case, the decimal point is being moved left 14 places, resulting in a solution of 8.016×10^{14}.

VERY SMALL NUMBERS

When dealing with a very small number, the decimal point is going to be moved to the right. A popular small measurement with biophysicists and transistor makers is the micron, which is one-millionth of a meter, or $.000001$ m. In scientific notation, that's 1.0×10^{-6}. Count the number of places to the right that the decimal point was moved.

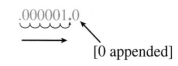

[0 appended]

The 0 is appended on the right of the first part of the numeral to justify the decimal point's being there, since a decimal point is never written as the last mark to the right of a numeral. Notice that when the decimal point is moved to the right, the 10 multiplier's exponent is negative.

Try two on your own. Express 0.000000008 and 0.0000000000358 in scientific notation. No peeking!

Here are the solutions: The first is 8.0×10^{-9}. I trust you weren't fooled by the 0 to the left of the decimal point. The second is 3.58×10^{-10}.

Practice Questions

1 $\dfrac{5}{8} - \left(-\dfrac{1}{3}\right) = $ _____

2 $-\dfrac{3}{4} + \left(-\dfrac{1}{6}\right) = $ _____

3 $\dfrac{7}{8} - \left(+\dfrac{1}{6}\right) = $ _____

4 $\dfrac{4}{9} \times \left(-\dfrac{2}{7}\right) = $ _____

5 $-\dfrac{4}{5} \times \dfrac{2}{3} = $ _____

6 $-\dfrac{7}{12} \div -\dfrac{3}{5} = $ _____

7 $+\dfrac{4}{5} \div \left(-\dfrac{3}{8}\right) = $ _____

8 $-5\dfrac{3}{8} \times -3\dfrac{3}{4} = $ _____

9 $-3\dfrac{4}{5} \div 2\dfrac{3}{4} = $ _____

Express each of the following fractions in its simplest form.

10 $\dfrac{-\dfrac{1}{2} - \left(-\dfrac{1}{3}\right)}{-\dfrac{2}{5}} = $ _____

11 $\dfrac{-\dfrac{2}{5} - \left(+\dfrac{5}{8}\right)}{-\dfrac{2}{5} + \left(-\dfrac{3}{8}\right)} = $ _____

Express each of the following decimals as a fraction in its simplest terms.

12 0.67   13 −0.10   14 −0.625

Express each of the following fractions as a decimal.

⑮ $\frac{3}{8}$

⑯ $-\frac{1}{3}$

⑰ $-\frac{7}{10}$

Express each of the following decimals as a percent.

⑱ 0.08

⑲ 0.675

⑳ 3.8725

Express each of the following percents as a decimal.

㉑ 35%

㉒ 7%

㉓ 165%

Solve each of the following.

㉔ 20% of 80

㉕ 15% of 200

㉖ A $250 suit is on sale for 45% off. What is the suit's sale price?

㉗ 50,000 people attended last Sunday's ballgame. 65% of them ate a hot dog during that game. How many hot dogs were eaten at last Sunday's game?

Express each of the following quantities in scientific notation.

㉘ 12,000,000

㉙ 302,000,000,000

㉚ 0.00000089

㉛ .00000000000006

Chapter Practice Answers

1 $\dfrac{5}{8} - \left(-\dfrac{1}{3}\right) = \dfrac{5}{8} + \left(\dfrac{1}{3}\right) = \dfrac{15}{24} + \dfrac{8}{24} = \dfrac{23}{24}$

2 $-\dfrac{3}{4} + \left(-\dfrac{1}{6}\right) = -\dfrac{9}{12} + \left(-\dfrac{2}{12}\right) = -\dfrac{11}{12}$

3 $\dfrac{7}{8} - \left(+\dfrac{1}{6}\right) = \dfrac{7}{8} + \left(-\dfrac{1}{6}\right) = \dfrac{21}{24} + \left(-\dfrac{4}{24}\right) = \dfrac{17}{24}$

4 $\dfrac{4}{9} \times \left(-\dfrac{2}{7}\right) = -\dfrac{8}{63}$

5 $-\dfrac{4}{5} \times \dfrac{2}{3} = -\dfrac{8}{15}$

6 $-\dfrac{7}{12} \div -\dfrac{3}{5} = -\dfrac{7}{12} \times -\dfrac{5}{3} = \dfrac{35}{36}$

7 $+\dfrac{4}{5} \div \left(-\dfrac{3}{8}\right) = +\dfrac{4}{5} \times \left(-\dfrac{8}{3}\right) = -\dfrac{32}{15} = -2\dfrac{2}{15}$

8 $-5\dfrac{3}{8} \times -3\dfrac{3}{4} = -\dfrac{43}{8} \times -\dfrac{15}{4} = \dfrac{645}{32} = 20\dfrac{5}{32}$

9 $-3\dfrac{4}{5} \div 2\dfrac{3}{4} = -\dfrac{19}{5} \div \dfrac{11}{4} = -\dfrac{19}{5} \times \dfrac{4}{11} = -\dfrac{76}{55} = -1\dfrac{21}{55}$

10 $\dfrac{-\dfrac{1}{2} - -\dfrac{1}{3}}{-\dfrac{2}{5}} = \dfrac{-\dfrac{1}{2} + +\dfrac{1}{3}}{-\dfrac{2}{5}} = \dfrac{-\dfrac{3}{6} + \left(+\dfrac{2}{6}\right)}{-\dfrac{2}{5}} = \dfrac{-\dfrac{1}{6}}{-\dfrac{2}{5}} = -\dfrac{1}{6} \times -\dfrac{5}{2} = +\dfrac{5}{12}$

11 $\dfrac{-\dfrac{2}{5} - \left(+\dfrac{5}{8}\right)}{-\dfrac{2}{5} + \left(-\dfrac{3}{8}\right)} = \dfrac{-\dfrac{2}{5} + \left(-\dfrac{5}{8}\right)}{-\dfrac{2}{5} + \left(-\dfrac{3}{8}\right)} = \dfrac{-\dfrac{16}{40} + \left(-\dfrac{25}{40}\right)}{-\dfrac{16}{40} + \left(-\dfrac{15}{40}\right)} = \dfrac{-\dfrac{41}{40}}{-\dfrac{31}{40}} = -\dfrac{\overset{7}{\cancel{41}}}{\underset{1}{\cancel{40}}} \times -\dfrac{\overset{1}{\cancel{40}}}{\underset{5}{\cancel{31}}} = \dfrac{41}{31} = 1\dfrac{10}{31}$

12 You have to recognize $0.67 = 0.6\overline{6} = \dfrac{66\overline{6}}{1000} = \dfrac{2}{3}$

⑬ $-0.10 = -\dfrac{10}{100} = -\dfrac{1}{10}$

⑭ $-0.625 = -\dfrac{625}{1000} = -\dfrac{5}{8}$

⑮ $\dfrac{3}{8} = 0.375$

⑯ $-\dfrac{1}{3} = -.33\overline{3}$ or $-.3\overline{3}$

⑰ $-\dfrac{7}{10} = -0.7$

⑱ $0.08 = 8\%$

⑲ $0.675 = 67\dfrac{1}{2}\%$ or 67.5%

⑳ $3.8725 = 387\dfrac{1}{4}\%$ or 387.25%

㉑ $35\% = 0.35$

㉒ $7\% = 0.07$

㉓ $165\% = 1.65$

㉔ $80 \times 0.2 = 16$

㉕ $200 \times 0.15 = 30$

㉖ There are two ways to solve this. One is to find 45% of $250 and subtract that amount from $250. The other is to think, "If the suit is discounted 45%, I'm paying 55% of $250." Either way, the sale price is $137.50.

㉗ 65% of $50,000 = 32,500$ hot dogs

㉘ $12,000,000 = 1.2 \times 10^{7}$

㉙ $302,000,000,000 = 3.02 \times 10^{11}$

㉚ $0.00000089 = 8.9 \times 10^{-7}$

㉛ $.00000000000006 = 6.0 \times 10^{-14}$

Variables, Terms, and Simple Equations

Every number that you have dealt with in the first three chapters of this book has been a constant—that is, it has a constant value. You've probably heard that *a rose is a rose is a rose*. Well one rose may be quite different from another rose in color, size, fragrance, and other ways. Far truer than that Gertrude Stein quote is *a three is a three is a three.* Be it three persons, three cans of tuna fish, three bowling pins, or three blue whales, each of them has a certain three-ness about it that can be represented by holding up three fingers on one hand. That's what makes 3 a constant. In this chapter, I introduce a new type of number called a **variable**. A variable has a temporary value in any situation. The same variable may stand for 5 in one problem and turn around and stand for 17 in another. Evaluating what number a variable is worth in any given situation is accomplished by solving equations. This chapter affords you the chance to become familiar with both of these.

Balance Pictures

You've probably seen an equal arm balance, whether in pictures representing the scales of justice or in a school science laboratory. Such a balance is represented here. Each pan of this balance holds a certain number of rolls of washers and a certain number of loose washers. Each roll contains an identical number of washers.

Notice that the balance is in **equilibrium**. That means that the amount of weight pushing down on each pan is the same. What you want to find out is how many washers are in each roll. If you count carefully, you'll see that there are 8 rolls and 6 loose washers on the left pan and 6 rolls and 24 loose washers on the right one. If you remove one washer at a time from both pans, you won't affect the equilibrium. In fact, as long as you do the same thing to both pans, the balance remains in equilibrium. (You could add the same amount to each pan as well, but that wouldn't suit your purposes here.)

Here, six loose washers have been removed from each side of the balance, leaving nothing but rolls of washers on the left side.

Suppose you start removing rolls from both sides until you can't remove any more without upsetting the equilibrium. That would take you to this situation.

From what remains, you should be able to determine how many washers are in each roll. Have you figured it out yet? There are 2 rolls on the left pan and 18 loose washers on the right. Dividing both of those quantities by 2, you can determine that one roll contains nine washers. I'm not sure how to break this to you, but you've just solved your first algebraic equation.

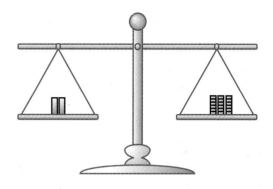

Constants and Variables

Defining Constants and Variables

Every number that you have ever dealt with before algebra has fallen into one category: **constants**. A constant is a number that has a specific value. Whether that value is **positive** or **negative**, **5** is always worth **5**, –7 is always worth –7, and so forth. Fractions and decimals are constants, too. A percent is a slightly different situation, because 5% of one number is not worth the same amount as 5% of another number; however, 5% of the same number is always the same, so a percent can also be considered a constant.

The type of number that's exclusive to algebra is the **variable**. A variable is a letter that takes the place of a number. The same variable—x, for example—may stand for one number on one occasion and a different number on another occasion. But every time a variable appears within the same problem, it must be worth the same amount.

The first example of a variable in this book is each roll of washers in the preceding section. There were nine washers in each roll. Within that problem, each roll of washers was worth the same thing—as much as nine loose washers. Of course, another time or in another problem, there might be a different number of washers in each variable.

FAQ

Which letters can be used as variables?

The number of constants is infinite—without end. The available number of variables in the American alphabet is only 26. How could the variables not be reused? In actuality, not all 26 letters are really available for use as variables. The letters i and e have special meanings and should not be used. It's also prudent not to use o, to avoid possible confusion with the number 0.

Combining Constants and Variables

Constants and variables, as well as variables and variables, can be combined in all the usual ways in which constants are combined with other constants. That said, the resulting combination won't necessarily make any sense.

There's addition:

$n + 3$
$4 + n$
$m + n$
$n + m$

What do they mean? Well, $n + 3 = n + 3$, $4 + n = 4 + n$. Do you get the idea? The fact that it's possible to write those combinations is one thing. They'll only have meaning if you know the numerical values for which m and n stand.

Continuing with the combinations, there's subtraction:

$n - 3$
$4 - n$
$m - n$
$n - m$

There's division: $n \div 3$, but more commonly $\frac{n}{3}, \frac{m}{n}, \frac{n}{4}$ and $\frac{n}{m}$.

Did you notice that I saved multiplication for last? Think it was an accident? It never is. Until now, I've used the $\times$ sign on more than one occasion, even though I mentioned earlier that it isn't used in algebra. That's because we haven't been doing algebra until now. The introduction of variables marks the disappearance of $\times$. I discussed other ways to indicate multiplication in Chapter 1. Here's where it begins.

$3n$
$4(n)$
mn
$n \cdot m$

Note that when the constant and the variable are written next to each other, as in $3n$, the constant always comes first. You will *never* see an expression like $x8$. When writing variables next to each other, as in mn, it is customary, but not essential, to write them in alphabetical order.

Coefficients and Factors

Factor is a word you've seen before, earlier in this book. Coefficient is probably new to you. Both words can mean the same thing, and in some contexts do. When written in the form $2x$, for example, 2 and x are both factors of each other and coefficients of each other. Don't let that confuse you. It's just the algebra devil at work!

COEFFICIENTS

When a constant and one or more variables are written next to each other to indicate multiplication, as is the case with the expressions $3w$ and $5mn$, the constant portion of each expression is known as the numeric **coefficient** of the variable(s). Each variable is also a coefficient of the other members of the expression. That is, w is a coefficient of 3; m and n are both coefficients of 5 and of each other. It is customary, however, for the term *coefficient* to be used only when referring to the numerical part of the expression.

Consider the variable x in the statement $x = 3$. What is x's coefficient? You might be tempted to say that it doesn't have one, or that it's 0. Well, you'd be incorrect in either case. If x's coefficient were 0, then the statement $x = 3$ could be rewritten as $0x = 3$, which would give the left side of the expression a value of 0. Remember, $0x$ means "0 times x," and 0 times anything equals 0. That turns the statement $0x = 3$ into $0 = 3$, which is clearly impossible. What, then, is x's coefficient? It's 1. Anytime you see a variable without a visible coefficient, it has a coefficient of 1 (understood), which brings up another peculiarity of algebra conventions. A coefficient of 1 is never written. You'll never see $1x$ or $1y$. You look at this again in a little while when you get to deal with evaluating expressions.

FACTORS

The word **factor** can be used in math as both a noun and a verb. As a noun, a factor is a number that is multiplied by another number to make a third number. For example:

- 3 and 2 are factors of 6.

- 1 is a factor of every natural number, since any natural number times 1 is itself.

- The factors of 12 are 1, 2, 3, 4, 6, and 12.

As a verb, to factor is to "unmultiply" something, or to break it into its component factors, while not changing its value. Often, there are many ways to factor a number. 12, for example, factors as $(1)(12)$, $(2)(6)$, and $(3)(4)$.

Variables can be used to change verbal expressions into algebraic ones. Here's a list of some:

Verbal Phrase	Algebraic Expression
8 more than a number	$n + 8$ or $8 + n$
the sum of a number and 5	$n + 5$ or $5 + n$
6 less than a number	$n - 6$
a number diminished by 5	$n - 5$
6 times a number	$6n$
4 more than the product of 8 and x	$8x + 4$

Evaluating Expressions

To evaluate an expression, simply replace the variable with its value and do the arithmetic. Sometimes it might be necessary to use grouping symbols as well.

If $z = 5$, find the values of:

a) $z + 7$

b) $12 - z$

c) $3z - 5$

> In the case of a): $z + 7 = 5 + 7 = 12$
> For b): $12 - z = 12 - 5 = 7$
> To solve c), you need grouping symbols: $3z - 5 = 3(5) - 5$
> $$15 - 5 = 10$$

Let's try three more. If $y = -4$, find the values of:

a) $y + 7$

b) $12 - y$

c) $4y - 7$

For these, you'll need grouping symbols for two of the three.

> For a) you get: $y + 7 = -4 + 7 = 3$
> For b): $12 - y = 12 - (-4)$
> $12 - (-4) = 12 + 4 = 16$
> Finally, for c): $4y - 7 = 4(-4) - 7$
> $4(-4) - 7 = -16 - 7 = -23$

An *equation* is a mathematical sentence. Like an English sentence, an equation is a complete thought. Unlike an English sentence, an equation is *always* a true statement. If a mathematical sentence does *not* express a true complete thought, it is *not* an equation. The verb in an equation is the equal sign. The next several sections deal with solving one-variable equations; that is—equations that contain only a single variable, such as x, y, or z.

Solving One-Variable Equations by Addition

You've seen equations and how to solve them before in this chapter, at least once. Do you remember the balance pictures at the very beginning of this chapter? An equation works exactly like an equal arm balance. The pivotal point of a balance is called its fulcrum. The = is an equation's fulcrum. If you add the same amount to both sides or remove the same amount from both sides, the equation remains in equilibrium.

EXAMPLE 1

The key to solving equations was just stated, but I can't stress it enough. Consider this equation:

$$x - 9 = 16$$

The main strategy for solving a one-variable equation is to get the variable by itself on one side of the equal sign. What is currently in the way of the variable being by itself? The 9, of course. How is the 9 combined with the variable? Currently, it is combined with it by subtraction. The minus sign should tell you that. How do you undo a subtraction?

$$\begin{array}{r} x - 9 = 16 \\ +9 \quad +9 \\ \hline \end{array}$$

The answer is to add. To undo a minus 9, add 9—but remember, whatever you do to one side of an equation must also be done to the other side to keep it in balance.

$$\begin{array}{r} x - 9 = 16 \\ +9 \quad +9 \\ \hline x \quad = 25 \end{array}$$

Completing the additions, $-9 + 9 = 0$, which you don't bother to write; $16 + 9 = 25$, so the equation is solved: $x = 25$.

EXAMPLE 2

Try finding the value of y in this equation.

$$y - 17 = 42$$

Go ahead and work it out on a piece of scrap paper, asking yourself appropriate questions as you sketch it out. Then, when you've done it, come back and see whether your solution agrees with mine.

$$\begin{array}{r} y - 17 = 42 \\ +17 \quad +17 \\ \hline y \quad\quad = 59 \end{array}$$

Solving One-Variable Equations by Subtraction

Before, we added the same amount to both sides to solve the equations while keeping them in equilibrium. These equations work a little differently, and that should be apparent as soon as you look at them.

EXAMPLE 1

Look at this equation and find out what's different about it.

$$21 = 7 + n$$

There are actually two main differences. The first is that the 7 and the n are combined by addition (remember, $7 + n$ and $n + 7$ mean the same thing). The second is that the variable is on the right side. Remember, an equation is like a balance. It makes no difference which side the weights are on, only that it is in equilibrium. Do you remember when we removed washers from each pan of the balance one at a time? Since the 7 is added to the variable and the way to undo an addition is by subtraction, that's what you do.

Remember, though, that when you subtract 7 from one side of the equation, you must do the same to the other side. That way you find that $n = 14$, which is the preferred way to leave the final answer, although $14 = n$ would not be incorrect.

$$\begin{array}{r} 21 = 7 + n \\ -7 \quad -7 \\ \hline 14 = \quad\quad n \end{array}$$

EXAMPLE 2

Try solving this more conventional-looking equation:

$$r + 11 = 8$$

. . . .

Did you solve it? Your solution should look like this.

If you didn't get $r = -3$, go over the steps again. You'll get a chance to practice in a bit.

$$
\begin{array}{rl}
r + 11 = & 8 \\
-11 & -11 \\
\hline
r \quad = & -3
\end{array}
$$

Solving One-Variable Equations by Division

Solving equations by multiplication is a subject I'll put off until I deal with ratio and proportion in Chapter 5. To solve by division, however, you need to understand that the principle is identical to solving by subtraction. After all, one definition of division, you may recall, is repeated subtraction of the same number. Although that justifies application of the same principle as solving by subtraction, you must keep sight of the fact that division is also defined as the undoing of multiplication.

EXAMPLE 1

Look at this equation:

$$3y = 15$$

. . . .

How are the 3 and the y combined? By now, you should know that they are coefficients of each other, meaning that they are connected by multiplication. How do you undo a multiplication? Like this:

$$3y = 15$$
$$\frac{3y}{3} = \frac{15}{3}$$
$$y = 5$$

As before, what you do to one side (divide by 3), you also do to the other, thus maintaining equilibrium.

EXAMPLE 2

Try one more. Find w in $-4w = 24$.

Remember, the objective is not to find $-w$, but to find w.

$$-4w = 24$$
$$\frac{-4w}{-4} = \frac{24}{-4}$$
$$w = -6$$

Solving One-Variable Equations Requiring Two Steps

Realistically, you are rarely going to see equations as simple as the ones that you've solved so far in this chapter. They do, however, form the basis for solving more complex equations. When solving an equation requires more than a single step, some combination of the steps used in the preceding sections will likely be used. In that case, it's good to remember that if division is going to be needed, it should be saved for last. Any additions or subtractions should precede dividing.

EXAMPLE 1

Let's try this one:

$$3k - 7 = 29$$

Since there's a 7 subtracted from the same side as the variable, you have to add 7 to get it off that side:

$$3k - 7 = 29$$
$$\underline{+7 \quad +7}$$
$$3k \quad\;\; = 36$$

Now the variable is combined with 3 by multiplication. You know what to do about that.

Divide both sides by 3 and find that $k = 12$.

$$3k = 36$$
$$\frac{3k}{3} = \frac{36}{3}$$
$$k = 12$$

EXAMPLE 2

Try one more example of an equation requiring two steps to solve.

$$7x + 8 = -27$$

This time, an 8 is added to the variable, so you're going to have to subtract 8 to get rid of it from that side:

$$7x + 8 = -27$$
$$\underline{-8 \quad -8}$$
$$7x \quad\;\; = -35$$

Now, it's time to deal with that 7 multiplier. And, there, you have it!

$$7x = -35$$
$$\frac{7x}{7} = \frac{-35}{7}$$
$$x = -5$$

Sure you can add and subtract variables. Soon you'll be able to add them in your sleep. In fact, the tendency to do so has caused me to wake up in the middle of the night, having added two variables I hadn't before thought possible, and yell "Eureka!" But that's a story for a different time and place. Besides, it's Greek to me!

ADDING VARIABLES

Until now, you've had no need to add variables, so I have refrained from going into the technique, but, to quote Lewis Carroll, "The time has come, the Walrus said, to speak of many things . . ." or, in this case, one specific thing. To be able to meaningfully add or subtract variables, *the variable portions of the expressions must be identical*. That is:

$$2r + 3r = 5r$$

The technique, as you can see, is to add the numeric coefficients and not change the variable. That should not be interpreted to mean that two different variables can't be added, because you've already seen that they can. For example, the expression $2v + 3w$ is perfectly permissible, but the result of the addition is as shown at right:

$$2v + 3w = 2v + 3w$$

Obviously, the rule of adding the numerical coefficients does not apply here.

SUBTRACTING VARIABLES

Subtracting variables works the same way, except that you subtract rather than add, so:

$$5n - 3n = 2n$$
$$7x - 3x = 4x$$
$$5y - 6y = -1y, \text{ which would be written } -y$$

Similarly:

$$2g - 3h = 2g - 3h$$

Perhaps not quite so obvious are the following situations:

$$5n - 4n = n$$
$$7x - 7x = 0$$
$$y + y = 2y$$

Does that make sense to you? You have to remember that y is really $1 \cdot y$, so $1y + 1y = 2y$. Also, $7x - 7x = 0x$, which is the same as $0 \cdot x$, which equals 0.

Changing Repeating Decimals to Fractions

Now for a brief change of pace. Changing repeating decimals to fractions would at first glance seem to belong back in Chapter 3, but in fact, you didn't have the knowledge at that time to have made any sense out of it. Now you do. Chapter 3 mentioned decimals that repeat infinitely, such as $0.\overline{3}$, which is 0.33333333, or $0.\overline{16}$, which stands for 0.161616.... Until now, there was no way to sensibly change one of these decimal fractions into a common fraction. Now that you know something about working with variables, however, you can do something about it.

Let d stand for $0.\overline{7}$ or $0.7777\ldots$.

Then $10d$ would be $7.\overline{7}$ or $7.7777\ldots$.

Since both d and $10d$ have the same decimal part, the difference between them must be an integer:

$$10d = 7.\overline{7}$$
$$-d = 0.7$$
$$9d = 7$$

If $9d = 7$, what does that make d? Simply divide both sides of the equation by 9 and get:

$$d = \frac{7}{9}$$

That means that $0.\overline{7} = \frac{7}{9}$. Cool, huh?

Try finding the fraction that means the same as the repeating decimal $0.\overline{16}$. Multiplying it by 10 isn't going to do it because the difference between $1.6\overline{16}$ and $0.\overline{16}$ is not an integer. However, if $n = 0.\overline{16}$, then $10n = 1.6\overline{16}$ and $100n = 16.\overline{16}$.

Now you have an integer difference between n and $100n$, since both have the same fractional part.

$$100n = 16.\overline{16}$$
$$-n = 0.16$$
$$99n = 16$$

Then, by the process of divide and conquer, you get $n = \frac{16}{99}$, so $0.\overline{16} = \frac{16}{99}$.

You've probably suspected that this moment was coming. I have covered all the principles you'll need to solve multiple-step equations, save for one.

EXAMPLE 1

Look at this equation.
$$9 - 3x + 11 = 16 - 2(3x - 8)$$

Variables and constants are on both sides of the equal sign, and parentheses need clearing. You learned back in Chapter 1 that clearing parentheses takes precedence. Following that, combine like terms if possible on either side of the equation and do whatever is necessary to *collect the terms* so that all the constants are on one side of the equation and all the variables are on the other. It's conventional to collect all the variables on the left, but that's entirely up to you. Finally, combine like terms again, if possible, and solve.

First you multiply what's in the parentheses by -2 (yes, it looks like a minus 2, but when you have to multiply by it, it gets treated like a -2):

$$9 - 3x + 11 = 16 - 2(3x - 8) \rightarrow 9 - 3x + 11 = 16 + -2(3x - 8)$$
$$9 - 3x + 11 = 16 - 6x + 16$$

Now combine the two constants on the left and combine the two constants on the right:

$$9 - 3x + 11 = 16 - 6x + 16$$
$$20 - 3x = 32 - 6x$$

It's starting to look easier, isn't it? You could now collect all your terms by adding things to both sides, but I'm going to illustrate it one step at a time, so let's get all the constants onto the right side by subtracting 20 from both sides:

$$20 - 3x = 32 - 6x$$
$$\underline{-20 \qquad\quad -20}$$
$$-3x = 12 - 6x$$

Next, collect the variables by adding $6x$ to both sides:

$$-3x = 12 - 6x$$
$$\underline{+6x \qquad + 6x}$$
$$3x = 12$$

Finally, divide both sides by 3.

That wasn't so bad, was it?

$$\frac{3x}{3} = \frac{12}{3}$$
$$x = 4$$

EXAMPLE 2

Try one more.

Why don't you try doing it on your own on scrap paper or in a notebook? Then come back and see whether your solution agrees with mine.

$$8 + 4y - 17 = 9 - 5y + 14 - 7y$$

First, combine like terms on each side of the equation:

$$8 + 4y - 17 = 9 - 5y + 14 - 7y$$
$$4y - 9 = 23 - 12y$$

Next, collect terms by adding 9 to both sides and adding $12y$ to both sides:

$$4y - 9 = 23 - 12y$$
$$\underline{+12y + 9 \quad + 9 + 12y}$$
$$16y \qquad = 32$$

Finally, divide:

$$\frac{16y}{16} = \frac{32}{16}$$
$$y = 2$$

Now I'll briefly complete the operations with variables topic before going on to practice.

MULTIPLYING VARIABLES

When two different variables are multiplied, the result is:

$$m \cdot n = mn$$

When two of the same variables are multiplied, as in $x \cdot x$, the base remains the same, but their exponents are added. For example, the preceding equation's product is x^2.

$$y \cdot y^2 = y^3$$
Finally, $n^2 \cdot n^3 = n^5$

When the variables have numeric coefficients, you multiply those coefficients in the usual fashion:

$$2n^2 \cdot 3n^3 = 6n^5$$

DIVIDING VARIABLES

When variables with the same base are divided, the base remains the same, but you subtract the exponents. If there are numeric coefficients, they are divided:

$$\frac{a^6}{a^2} = a^4; \; 6c^5 \div 2c^3 = 3c^2$$

Practice Questions

1 What is the difference between a constant and a variable?

2 In the expression $5n$, which is the coefficient?

3 If $w = 4$, find the value of $5w + 3w + 9$.

4 If $n = -3$, find the value of $4n - 3n - 9$.

Solve each of the following for the specified variable.

5 $x - 7 = 13$

6 $y + 11 = 19$

7 $5m = 45$

8 $3t - 11 = 7$

9 $5p + 11 = 41$

10 $3r - 9 = 18 - 6r$

11 $8 - 5x = 7x - 16$

12 $7 - 4v + 6 = -3v + 29 - 5v$

Write each of the following as a fraction.

13 $0.\overline{8}$

14 $0.\overline{51}$

15 $0.\overline{682}$

Chapter Practice Answers

1 A constant has a fixed value, such as 3, 5, or 27. A variable can stand for one amount in one situation and a different amount in another.

2 5 is the numerical coefficient of n, but both 5 and n are coefficients of each other.

3 $5 \cdot 4 + 3 \cdot 4 + 9 = 20 + 12 + 9 = 41$

4 $4(-3) - 3(-3) - 9 = -12 + 9 - 9 = -12$

5 $x - 7 = 13$

$$\underline{+7 \quad +7}$$

$x \quad = 20$

6 $y + 11 = 19$

$$\underline{-11 \quad -11}$$

$y \quad = \quad 8$

7 $\dfrac{5m}{5} = \dfrac{45}{5}$

$m = 9$

8 $3t - 11 = \quad 7$

$$\underline{+11 \quad +11}$$

$3t \qquad = 18$

$\dfrac{3t}{3} = \dfrac{18}{3}$

$t = 6$

9 $5p + 11 = 41$

$$\underline{-11 \quad -11}$$

$5p \qquad = 30$

$\dfrac{5p}{5} = \dfrac{30}{5}$

$p = 6$

⑩ $3r - 9 = 18 - 6r$

$\underline{+6r + 9 \quad +9 + 6r}$

$9r \qquad = 27$

$\dfrac{9r}{9} = \dfrac{27}{9}$

$r = 3$

⑪ $8 - 5x = 7x - 16$

$\underline{-8 - 7x \quad -7x - 8}$

$-12x = -24$

$\dfrac{-12x}{12} = \dfrac{-24}{12}$

$x = 2$

⑫ $7 - 4v + 6 = -3v + 29 - 5v$

$13 - 4v \qquad = 29 - 8v$

$\underline{-13 + 8v \qquad -13 + 8v}$

$4v \qquad = 16$

$\dfrac{4v}{4} = \dfrac{16}{4}$

$v = 4$

⑬ $n = 0.\overline{8}$, so $10n = 8.\overline{8}$

$9n = 10n - n = 8.\overline{8} - 0.\overline{8}$

$9n = 8, \therefore n = \dfrac{8}{9}$

⑭ $n = 0.\overline{51}$, so $100n = 51.\overline{51}$

$99n = 100n - n = 51.\overline{51} - 0.\overline{51}$

$99n = 51, \therefore n = \dfrac{51}{99}$

⑮ $d = 0.\overline{682}$, so $1000d = 682.\overline{682}$

$999d = 1000d - d$

$999d = 682.\overline{682} - 0.\overline{682}$

$999d = 682 \therefore d = \dfrac{682}{999}$

$^*\therefore$ means therefore.

chapter 5

Axioms, Ratios, Proportions, and Sets

This chapter begins with axioms of equality, which are basic truths that everyone assumes are so without ever needing to prove them. From there, the discussion moves on to ratios that are algebraic comparisons between two different quantities. In order to learn anything from ratios, it is necessary to equate one comparison with another of equal value. Such an equation is called a **proportion**, which can be solved through the use of cross multiplication. Just when you may think that I'm getting ready to blow everything way out of proportion, get ready, set. . . . If you're waiting for "go," forget about it. Set is the thing you deal with next—set theory that is. After examining special sets and types and how to describe them, this chapter concludes with the union and intersection—two different ways of joining sets.

Axioms of Equality

An axiom is a logical statement or rule that is accepted as true without the need for proof, since it is inherently obvious. For all real numbers a, b, c, and d, these are some basic rules that govern the way in which you use the equal sign.

AXIOM OF REFLEXIVITY

The axiom of reflexivity, also known as the reflexive axiom, states that any number is equal to itself: $a = a$.

This applies to variables *only within a single equation or set of equations*. Reflexivity applies to constants *always*. $6 = 6$ at all times.

AXIOM OF SYMMETRY

The axiom of symmetry, also known as the symmetric axiom, states that if one number is equal to a second number, then the second number is equal to the first number. Stated with variables, if $a = b$ then $b = a$.

A practical application of this is if $3 + 5 = 8$, then $8 = 3 + 5$.

AXIOM OF TRANSITIVITY

The axiom of transitivity, also known as the transitive axiom, states that if one number is equal to a second number, and the second number is equal to the third number, then the first number equals the third number.

That is if $a = b$ and $b = c$, then $a = c$.

Extending that with constants, if $3 + 3 = 6$ and $6 = 4 + 2$, then $3 + 3 = 4 + 2$.

ADDITIVE AXIOM

The **additive axiom** states that if the first number is equal to a second number and a third number is equal to a fourth number, then the sum of the first number and the third number is equal to the sum of the second number and fourth number. Maybe you'd better read that again, because it's a mouthful.

Stated with variables, if $a = b$ and $c = d$, then $a + c = b + d$.

Perhaps that's better shown with constants where if $2 + 2 = 4$ and $3 + 3 = 6$, then $2 + 2 + 3 + 3 = 4 + 6$.

MULTIPLICATIVE AXIOM

The **multiplicative axiom** states that if the first number is equal to a second number and a third number is equal to a fourth number, then the product of the first number and the third number is equal to the product of the second number and fourth number.

Stated with variables, if $a = b$ and $c = d$, then $(a)(c) = (b)(d)$.

Illustrating that with constants, if $2 = 2$ and $3 = 3$, then $(2)(3) = (2)(3)$.

Neither of the last two axioms is very profound.

Defining and Creating Ratios

A ratio is a comparison of two different things. Ratios are used in arithmetic, algebra, and geometry. Ratios may be written using a colon, as in *a:b*, or as a fraction, *a/b* or $\frac{a}{b}$. Whichever way you write a ratio, it is read as "the first quantity (or the top of the fraction) is to the second quantity (or the bottom of the fraction). If that sounds like an incomplete statement, it is, since to complete it would require another part, namely ". . . as a third quantity is to a fourth quantity." The "as" is provided by an equal sign; the third and fourth quantities are provided by a second ratio. The two ratios with the "=" between them form what is known as a proportion. You will work with proportions in the next section.

When it comes to ratios, order matters. *a:b* and *b:a* are two different relationships—in fact, each is the reciprocal of the other. If Jake has five toy cars and Alex has six toy cars, the ratio of Jake's toy cars to Alex's is 5:6. The ratio of Alex's toy cars to Jake's is 6:5.

Reese drank three glasses of juice at lunch yesterday; Myles drank five glasses of juice at yesterday's lunch. How does the amount of juice Myles drank at lunch yesterday compare with what Reese drank? Hopefully, you are prepared to answer that question with $\frac{5}{3}$. And, if the question had been "how did Reese's intake of juice compare with Myles'," you'd have been prepared to answer $\frac{3}{5}$.

Proportion is defined in the section dealing with ratios, earlier. As noted, a proportion is an equating of two ratios. So, 50:70 is 5:7 is a proportion. So is $\frac{9}{12} = \frac{3}{4}$.

MEANS AND EXTREMES

Special names are associated with the parts of a proportion. When the ratios are written in colon form, the inner two numbers—those closest to the equal sign—are called **the means**. Those farthest from the equal sign are **the extremes**. You can see them in this figure.

$$\text{MEANS} \\ \downarrow \quad \downarrow \\ 6 : 5 = 18 : 15 \\ \uparrow \qquad \qquad \uparrow \\ \text{EXTREMES}$$

When the ratios are written in fraction form, you have to work a little harder to figure out which are the means and which are the extremes, as you can see in this figure.

$$\begin{array}{c} \text{EXTREMES} \\ \dfrac{9}{24} = \dfrac{3}{8} \\ \text{MEANS} \end{array}$$

A very useful rule is used to solve algebraic proportions. That rule is "the product of the means equals the product of the extremes." You'll see how that works in the next section.

SOLVING PROPORTIONS BY CROSS PRODUCTS

The way you solve proportions is the model for solving all algebra problems when faced with a fraction on either side of the equal sign. Since the product of the means equals the product of the extremes, as you've just learned, then the multiplication across the "=" has to be true, and, as you can see in this figure, is true.

$$\frac{3}{5} \times \frac{9}{15}$$

$$45 = 45$$

Cross product

But it's not just true in this case. If an equation is a true statement, which it must always be, then multiplying across the equal sign, as seen in this figure, must always be true, since a ratio may always represent a fraction, and vice versa. That means that anytime just a single fraction is on either side of the equation, that equation must be a proportion and soluble by **a cross product**, which is so-named for reasons that should be apparent from this figure.

"So what does this have to do with the price of beans?" you may well ask. Here's a problem that hopefully will clear that up. Kidd Stadium, a Little League park, seats 2/5s the number of people as the original Yankee Stadium, which held 70,000. How many people can Kidd Stadium seat? The proportion in words would be Kidd Stadium's seating is to Yankee Stadium's seating as 2 is to 5:

$$\frac{\text{Kidd Stadium's Seating}}{\text{Yankee Stadium's Seating}} = \frac{2}{5}$$

Let n stand for the number of people that Kidd Stadium can hold. You already know how many Yankee Stadium could hold, so the proportion becomes:

$$\frac{n}{70,000} = \frac{2}{5}$$

Now cross-multiply:

$$\frac{n}{70,000} \bowtie \frac{2}{5}$$

You know how to solve the rest:

$$5n = 2 \cdot 70,000$$
$$5n = 140,000$$
$$\frac{5n}{5} = \frac{140,000}{5}$$
$$n = 28,000$$

Here's another problem requiring solution by proportions.

Jason and Dylan were both collectors of exotic matchbooks. Jason has 7/12 as many matchbooks as Dylan. If Dylan has 3456 matchbooks, how many does Jason have? Begin by establishing the proportion:

$$\frac{\text{Jason's matchbooks}}{\text{Dylan's matchbooks}} = \frac{7}{12}$$

Next, pick a variable to represent the one quantity you don't know—
Jason's matchbooks; how about j? That makes your proportion:

$$\frac{j}{3456} = \frac{7}{12}$$

What do you suppose the next step is?

$$\frac{j}{3456} \diagdown\kern-1.2em\diagup \frac{7}{12}$$

You said "cross-multiply," no doubt. From there, you should be
familiar with how to solve the equation:

That means that Jason has 2016 matchbooks.

$$12j = 7 \cdot 3456$$
$$12j = 24{,}192$$
$$\frac{12j}{12} = \frac{24{,}192}{12}$$
$$j = 2016$$

Literal Proportions

Sometimes equations are made up only of letters. They are known as **literal** equations. Here, you deal
only with a couple of literal proportions.

Solve this proportion for x:

$$\frac{x}{d} = \frac{w}{t}$$

Other than the fact that this proportion has no constants, nothing is unusual about how you solve it. The
variable x is red to remind you that that's what you're solving for.

For openers, cross-multiply and get:

$$tx = dw$$

Next, isolate the desired variable by dividing both sides by the
quantity, t:

$$\frac{tx}{t} = \frac{dw}{t}$$

That's it. Nothing further can be done. It is solved for x.

$$x = \frac{dw}{t}$$

A set is a group of objects, numbers, symbols, and so on. It is usually named by a single uppercase letter and indicated by the use of braces {}. So $A = \{a, b, c, d\}$ is read "Set A is the set containing the elements a, b, c, and d." A member of a set is known as an element of it. $a \in$ Set A **is read "a is an element of Set A."**

SPECIAL SETS

A part of a set is called a subset of that set.

$$\{a, b\} \subset \{a, b, c, d\}$$

Actually a set that contains 4 elements has 16 possible subsets, namely:

Subsets taken 1 element at a time:

$\{a\}$ $\{b\}$ $\{c\}$ $\{d\}$

Subsets taken 2 elements at a time:

$\{a, b\}$ $\{a, c\}$ $\{a, d\}$ $\{b, c\}$ $\{b, d\}$ $\{c, d\}$

Subsets taken 3 elements at a time:

$\{a, b, c\}$ $\{a, b, d\}$ $\{b, c, d\}$ $\{a, c, d\}$

Subsets taken 4 elements at a time:

$\{a, b, c, d\}$ That's right; a set is a subset of itself. I'll bet that surprised you, but how about this last one?

Subsets taken 0 elements at a time:

$\{\}$ That's known as the empty set, or null set, and is symbolized by the Greek letter ϕ (phi).

The universal set is the broad category that all elements of a set are part of. For every set shown so far on this page, the universal set is lowercase letters of the American alphabet. For the set $\{c, d, 2, 3\}$, the universal set is lowercase letters and whole numbers.

FACT

Since the order of the elements does not matter, $\{a, b, d\}$, $\{b, a, d\}$, and $\{b, d, a\}$ are considered to be three forms of the same set, also known as equal sets.

DESCRIBING SETS

Sets can be described by any of three different methods. First, a set may be described by the **rule** it must follow. For example:

$$\{v \mid v < 97, v \text{ is a positive integer}\}$$

That is read as "the set of v such that v is less than 97 and is a positive integer."

Another rule to describe a set might be {all students in the school with red backpacks}

The second way to describe a set is by **roster**; that is by listing its members. Here are two examples:

$$\{4, 5, 6, \ldots\}$$
$$\{\text{Kira, Reese, Rocio, Hailee}\}$$

The third method for describing sets is pictorially, using **Venn diagrams**. You can look at these more thoroughly in a bit, when you deal with union and intersection, but here's a simple one:

A Venn Diagram

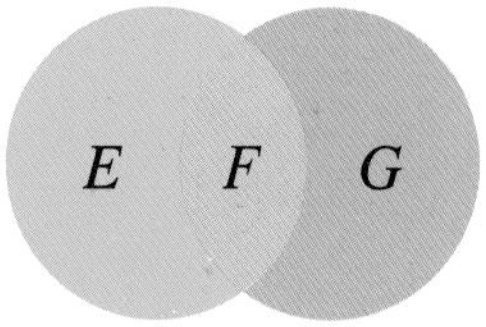

Notice that F is the intersection of the two circles. More about that in a bit.

All sets fall into one of two types—**finite** or **infinite**. A finite set has a definite number of elements; those elements can be counted and have a definite ending point.

$$B = \{1, 2, 3, 4\} = \{4, 3, 2, 1\}$$

These sets shown here are finite.

An infinite set's elements are incapable of being counted, since they go on forever, without stopping.

Here's an example of an infinite set:

$$C = \{1, 2, 3, 4, \ldots\}$$

TIP

Don't be fooled by an **ellipsis** of three dots (. . .). This set looks to be infinite, but it's not!

$$D = \{a, b, c, d, \ldots\}$$

Set D appears to be the same as Set C, only with letters, and there's the problem. The set of counting numbers (Set C) is infinite; the set of lowercase letters (Set D) ends with "z."

EQUAL VERSUS EQUIVALENT SETS

As already mentioned, **equal sets** are sets containing the exact same elements. Sets F and G are equal sets.

Set $F = \{e, f, g, h\}$
Set $G = \{g, f, h, e\}$

Equivalent sets is the name given to sets that contain the same number of elements. They are sometimes referred to as sets whose elements may be placed into a *one-to-one correspondence*.

Sets K and L are equivalent sets. The double-headed arrows indicate their one-to-one correspondence, although you could have just counted the number of elements in each.

Set $L = \{e, f, g, h\}$
$\updownarrow \quad \updownarrow \quad \updownarrow \quad \updownarrow$
Set $K = \{4, 5, 6, 7\}$

UNION AND INTERSECTION

The union of two or more sets is a set containing all the members of those sets. Here is a figure showing the union of two sets:

$$A = \{4,5,6\}$$
$$B = \{6,7,8\}$$
$$A \cup B = \{4,5,6,7,8\}$$

Notice that even though 6 appears in both sets A and B, which might lead you to expect that there would be one more element in the union than is actually present, *an element is never repeated* in a set. That does not mean, though, that the 6 is lost; in fact, it's given its own special place. It is known as the **intersection** of sets A and B and is represented as follows:

$$A \cap B = \{6\}$$

Consider the following situation:

In this case, the intersection had two elements.

$$C = \{a,b,c,d\}$$
$$D = \{c,d,e,f\}$$
$$C \cap D = \{c,d\}$$

Earlier in this chapter, I promised a closer look at Venn diagrams, so here you go. The figure to the right illustrates the intersection of sets C and D. Notice where $C \cap D = \{c,d\}$ appears.

Venn diagrams are also very useful for solving certain types of problems. Two examples follow.

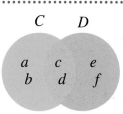

Eighty students participate in one or more of three sports: baseball, tennis, and basketball. Four students participate in all three sports; five play both baseball and basketball, only; two play both tennis and basketball, only; and three play both baseball and tennis, only. If seven students play only tennis and one plays only basketball, what is the total number of students who play only baseball?

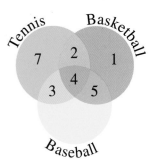

Check out the diagram. 22 of 80 students have been accounted for. 80 − 22 = 58 who play baseball only.

Here comes one more.

There are 75 students in a music program. Of those students, many either play in the school band or orchestra or sing in the chorus. Two of the students do all three. Fifteen of the students play in the band but don't sing in the chorus or play in the orchestra. Sixteen of the students sing in the chorus and play in the band. Twelve students sing in the chorus but don't play in the band or orchestra. Nine play in the band and in the orchestra only. Two play in the orchestra and sing in the chorus only. How many students play in the orchestra only?

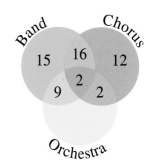

56 of the 75 students are accounted for on the diagram. Study what each section means. 75 − 56 = 19 in the orchestra only. They must be violinists!

Practice Questions

Fill in the blank to complete the specified axiom of equality.

❶ The axiom of reflexivity can be demonstrated by the expression $c =$ ___.

❷ The axiom of transitivity states that if $a = b$ and b = c, then $a =$ ___.

❸ The axiom of symmetry states that if $a = b$, then $b =$ ___.

❹ To illustrate the additive axiom, if $a = b$ and $c = d$, then $a + c =$ ___.

❺ The multiplicative axiom of equality may be illustrated by the expression if $a = b$ and $c = d$, then ___.

Solve each of the following.

❻ Jim ate two pieces of French toast for breakfast; Ian ate three pieces of French toast for breakfast.

   a) Compare Jim's intake of French toast to Ian's as a ratio.

   b) Compare Ian's intake of French toast to Jim's as a ratio.

   c) What portion of French toast did Jim eat?

❼ Ali spent $65 on party supplies for her birthday. Her husband Joe spent $95 for the same.

   a) Compare Ali's party supply expenditures to Joe's.

   b) Compare Joe's party supply expenditures to Ali's.

   c) Compare Joe's party supply expenditures to their total party supply expenditures.

Solve each for the variable.

❽ $\dfrac{6}{n} = \dfrac{4}{7}$

❾ $\dfrac{9}{6} = \dfrac{x}{8}$

⑩ $\dfrac{y}{15} = \dfrac{8}{3}$

⑪ $\dfrac{4}{7} = \dfrac{12}{p}$

Solve for either x or y.

⑫ $\dfrac{t}{r} = \dfrac{v}{x}$

⑬ $\dfrac{m}{y} = \dfrac{s}{h}$

Write a proportion that could be used to solve the problem. Then solve the problem.

⑭ The ratio of Tania's insurance bill to Dylan's was 7:5. Tania's bill was $14 more than Dylan's bill. What was Tania's insurance bill?

⑮ Rocio's hamburger stand sells three-eighths the number of hamburgers as the original Yoda's Hot Dog Stand sells hot dogs. If Yoda's sells 2000 hot dogs, how many hamburgers did Rocio's sell?

⑯ Marge and Karen both collect crackle-glass items. Karen has 342 of them. If that's only two-fifths the crackle-glass items that Marge has, how many crackle-glass items are in Marge's collection?

The following problems refer to sets. Bear that in mind when answering.

⑰ Write all subsets of $A = \{1,2,3\}$.

⑱ Set $B = \{m, n, o, p\}$. Set C is equivalent to B. Write Set C.

⑲ Write a set equal to $\{a, b, c, d, e\}$.

⑳ $D = \{1, 3, 5, 7, 9\}$, $E = \{5, 7, 9, 11, 13\}$.

   a) Write $D \cup E$.

   b) Write $D \cap E$.

Solve with the assistance of a Venn diagram.

21 In Ms. Willis's math class, eighteen students sing in the chorus, nine students are in band, five students participate in both activities, and seven students are in neither band nor chorus. How many students are in Ms. Willis's math class?

22 Fifty-five students drink one or more of three juices: cranberry, apple, and orange. Six students drink all three juices; seven drink both cranberry and apple only; four drink both apple and orange only; and five drink both orange and cranberry only. If nine students drink only cranberry and two drink only apple, what is the total number of students who drink only orange juice?

Chapter Practice Answers

1 $c = c$

2 $a = c$

3 $b = a$

4 $b + d$

5 $(a)(c) = (b)(d)$

6 a) $\frac{2}{3}$ or 2:3, Jim's 2 compared to Ian's 3

   b) $\frac{3}{2}$ or 3:2, Ian's 3 compared to Jim's 2

   c) $\frac{2}{5}$ or 2:5, Jim's 2 compared to the total number of pieces eaten

7 a) $\frac{65}{95}$ or $\frac{13}{19}$ or 65:95 or 13:19   Ali's $65 compared to Joe's $95

   b) $\frac{95}{65}$ or $\frac{19}{13}$ or 95:65 or 19:13   Joe's $95 compared to Ali's $65

   c) $\frac{95}{160}$ or $\frac{19}{32}$ or 95:160 or 19:32   Together they spent $95 + $65 = $160.

8 By cross-multiplying you get $4n = 42$. Then,

$$\frac{4n}{4} = \frac{21}{4}$$

$$n = 5.25 \text{ or } 5\frac{1}{4}$$

9 Cross-multiplying, you get $6x = 72$. Then,

$$\frac{6x}{6} = \frac{72}{6}$$

$$x = 12$$

10 By cross-multiplying you get $3y = 120$. Then,

$$\frac{3y}{3} = \frac{120}{3}$$

$$y = 40$$

⑪ Cross-multiplying gives $4p = 84$. Then,

$$\frac{4p}{4} = \frac{84}{4}$$
$$p = 21$$

⑫ Cross-multiplying gives $tx = rv$. Then,

$$\frac{tx}{t} = \frac{rv}{t}$$
$$x = \frac{rv}{t}$$

⑬ Cross-multiplying gives $sy = hm$. Then,

$$\frac{sy}{s} = \frac{hm}{s}$$
$$y = \frac{hm}{s}$$

⑭ Let x = Tania's insurance bill. Then Dylan's bill is $x - 14$. First write the proportion:

$$\frac{\text{Tania's bill}}{\text{Dylan's bill}} = \frac{7}{5}$$
$$\frac{x}{x - 14} = \frac{7}{5}$$

Next, cross-multiply and solve:

$$5x = 7(x - 14)$$
$$5x = 7x - 98$$
$$\underline{-7x - 7x}$$
$$-2x = \quad -98$$
$$\frac{-2x}{-2} = \frac{-98}{-2}$$
$$x = 49$$

Tania's bill was $49.

⑮ Let x = Rocio's hamburgers. Now set up the proportion:

$$\frac{x}{2000} = \frac{3}{8}$$

Now cross-multiply and solve:

$$8x = 6000$$
$$\frac{8x}{8} = \frac{6000}{8}$$
$$x = 750$$

That's 750 hamburgers.

⑯ Let n = the number of items in Marge's collection. The proportion is:

$$\frac{342}{n} = \frac{2}{5}$$

Remember, Karen has the smaller collection. Now cross-multiply and solve:

$$2n = 5(342)$$
$$\frac{2n}{2} = \frac{1710}{2}$$
$$n = 855$$

Marge has 855 crackle-glass items.

⑰ $\{1\}, \{2\}, \{3\}, \{1, 2\}, \{1, 3\}, \{2, 3\}, \{\}, \{1, 2, 3\}$.

⑱ Any set containing exactly four elements, e.g., $\{1, 2, 3, 4\}$.

⑲ Any arrangement of the letters $\{a, b, c, d, e\}$, regardless of their order, with each used once only.

⑳ a) $D \cup E = \{1, 3, 5, 7, 9, 11, 13\}$

b) $D \cap E = \{5, 7, 9\}$

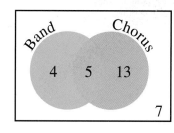

㉑ The rectangle is the universe, a.k.a. Ms. Willis's math class. Two circles are needed, which are labeled band and chorus. First, put the 5 into the intersection, since they take both band and chorus. The problem tells you that 18 are in the chorus, but 5 of those have already been accounted for, so another 13 are needed. 9 students are in the band, but 5 of those have already been accounted for (those in the intersection), so another 4 are needed. Finally, those who take neither are shown in neither circle, but complete the universe (7). Now add them up. $4 + 5 + 13 + 7 = 29$.

㉒ Don't look below this line until you've worked it out for yourself.

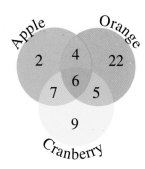

Add the ones on the diagram before the orange-juice-only number was added and you'll get 33. $55 - 33 = 22$, the number that drink orange juice only.

chapter 6

Monomials, Binomials, and Systems of Equations

Monomials and binomials are expressions containing one and two terms, respectively. I introduced the expression *term* in Chapter 4, but if you're not sure of its meaning, you can check the Glossary on p. 278. Mono- means *one*, as in "my wife must remain monogamous" (that'll teach her) or "monoplane," which is an airplane containing a single wing, as all modern ones do. Bi- means *two*, as in bipolar (as the earth) or biplanes, which you barely see anymore except at air shows. Monomials may contain a single term, but that single term can contain many parts, while binomials contain two terms separated by a plus or minus sign.

The rest of the chapter is largely concerned with operations involving monomials and binomials. The chapter builds to a crescendo with the solving of systems of equations in two variables by addition and subtraction as well as by the method known as substitution.

A **monomial** is an algebraic expression that consists of only a single term. A term may be a constant, a variable, or a combination of both. $5a$, $4b^2$, and $6x^3y^4$ are all examples of monomials. The key element to determining a monomial is the absence of a "+" or "–" sign.

ADDING AND SUBTRACTING MONOMIALS

When adding or subtracting monomials, the numerical portions are added or subtracted in the same way as are signed numbers, which is covered in Chapter 2. The variables, however, are not changed.

$$
\begin{array}{ll}
\text{a)} \quad \begin{array}{r} 5n \\ +5n \\ \hline 10n \end{array} & \text{b)} \quad \begin{array}{r} -5a^2b \\ +3a^2b \\ \hline 2a^2b \end{array}
\end{array}
$$

c) $3x^4y^2 - (-3x^4y^2) = 6x^4y^2$

To combine monomials by addition and/or subtraction, their variable portions must be identical. Unlike the three preceding examples:

$$3ab^2 + 2a^2b = 3ab^2 + 2a^2b$$

MULTIPLYING MONOMIALS

You looked at the rules for multiplying and dividing monomials in Chapter 1, but you only got a glimpse of multiplying variables. The definitions and rules for exponents and powers apply to monomials as well.

$$5 \cdot 5 = 5^2 = 25 \text{ and } x \cdot x \cdot x = x^3$$

In the same way,

$$a \cdot a \cdot a \cdot a \cdot b \cdot b \cdot b = a^4b^3$$

In order to multiply monomials containing variable and constant components, the variables of each type's exponents are added together; the constants are multiplied. Here are some examples:

Notice in the second example that the exponents of the as were added together and the exponents of the bs were added together.

$$n^4 \cdot n^5 = n^9$$

$$2a^3b^4 \cdot 4a^2b^3 = 8a^5b^7$$

The same pattern is followed here with the exponents of j, k, and l.

$$-5j^2k^3l^4 \cdot -6j^2k^3l^4 = 30j^4k^6l^8$$

Here's one more example:

$$4abc \cdot 7cde = 28abc^2de$$

Notice that only one variable, c, is common to both monomials.

RAISING MONOMIALS TO A POWER

In order to raise a monomial to a power, multiply each part of the monomial by the number to which it is being raised. Here are some examples:

$$(x^4)^3 = x^{12} \quad (3b^2c^3)^2 = (3)^2b^4c^6 = 9b^4c^6$$
$$(-2a^3b^4d^5)^3 = (-2)^3a^9b^{12}d^{15}$$
$$= (-2)(-2)(-2)a^9b^{12}d^{15}$$
$$= -8a^9b^{12}d^{15}$$

DIVIDING MONOMIALS

As with signed numbers, you're most likely to see division represented in fractional notation. The fraction to the right could be read as "24efg over 8ef" or as "24efg divided by 8ef."

$$\frac{24efg}{8ef}$$

It doesn't matter. Either way, to express it in simplest form requires dividing, which entails actual division of the top numerical portion by the bottom numerical portion, and the subtraction of the exponents of the variables on the bottom from those on the top. For that subtraction, you'll have to use a little imagination. That imagination causes you to rewrite the division in a form you'll never ever actually see.

Nobody ever writes anything raised to the power of 1, so you have to imagine that, but by doing so, you can see how dividing monomials really works. While the first fraction in the explanation shows no g in the denominator, there is a g in the denominator of the second fraction. Since anything raised to the 0 power has a value of 1, you can put anything you like in there raised to the 0 power. I chose to put the g there to clarify the rest of the explanation.

Finishing Line 1, the numeric from the top has been divided by that on the bottom to get 3, and each bottom variable's exponent has been subtracted from the appropriate top one. So, there are two variables raised to the 0 power and one raised to the power 1 (Line 2). To begin Line 3, multiplication dots have been inserted to show what's really going on; then, in the middle I've exchanged two of the variables for their real values and removed the exponent from the last. Finally, everything is multiplied together to come up with the quotient, $3g$.

Line 1 $\quad \dfrac{24efg}{8ef} \rightarrow \dfrac{24e^1 f^1 g^1}{8e^1 f^1 g^0} \rightarrow 3e^{1-1} f^{1-1} g^{1-0}$

Line 2 $\qquad 3e^{1-1} f^{1-1} g^{1-0} = 3e^0 f^0 g^1$

Line 3 $\quad 3 \cdot e^0 \cdot f^0 \cdot g^1 = 3 \cdot 1 \cdot 1 \cdot g = 3g$

Binomials

The prefix *bi-* means *two,* so you might readily interpret the meaning of binomials to be expressions containing *two terms,* separated by a plus or minus sign—the two signs *not allowed* in a monomial. Binomials are the smallest example of polynomials—expressions containing *many terms.* Two examples of binomials are $ax + b$ and $3y - 7$.

ADDING AND SUBTRACTING BINOMIALS

Adding or subtracting binomials may result in a monomial, another binomial, or a polynomial. That might sound strange at first, but these examples should clarify it for you.

$$(5y + 3) - (2y + 3) = 5y + 3 - 2y - 3$$
$$= 3y$$

That subtraction (above right) resulted in a monomial; next:

$$(5x + 7) + (2x + 1) = 7x + 8$$

That addition resulted in a binomial; finally (fanfare, please):

$$(4z^2 + 3z) - (5z - 8) = 4z^2 + 3z - 5z + 8$$
$$= 4z^2 - 2z + 8$$

And the subtraction resulted in a polynomial. The defense rests.

MULTIPLYING BINOMIALS

Binomials may be multiplied by stacking them and multiplying each term of the top one by each term of the bottom one, just as you might do in a multiplication of a two-digit number by another two-digit number, as follows. First multiply by the 3:

$$\begin{array}{r} 2x + 4 \\ \times\ x + 3 \\ \hline 6x + 12 \end{array}$$

Then multiply by the x and add:

$$\begin{array}{r} 2x + 4 \\ \times\quad x + 3 \\ \hline 6x + 12 \\ 2x^2 + 4x \\ \hline 2x^2 + 10x + 12 \end{array}$$

Notice that like variables are lined up for addition. Also notice that the dreaded × sign was used. This is at best a cumbersome operation, and at worst has the potential for confusion. Before you ask, the answer is, "Yes, there's a better way," and it's known as the **F.O.I.L.**, or simply the FOIL method. The letters stand for First Outer Inner Last. Here's the multiplication that you just did:

$$(2x + 4)(x + 3)$$

Consider each set of parentheses both separately and together and multiply, thusly:

First:	$(2x + 4)(x + 3) = 2x^2$
Outer:	$(2x + 4)(x + 3) = 6x$
Inner:	$(2x + 4)(x + 3) = 4x$
Last:	$(2x + 4)(x + 3) = 12$

Look down the final column and get $2x^2 + 6x + 4x + 12$. The two middle terms can, of course, be combined, and the result is $2x^2 + 10x + 12$. Does that product look familiar? It may sound like a bit much to ask you to memorize FOIL, and what it stands for, but I assure you that after you have, you'll find that it was well worth the time spent doing so in the time you'll save when multiplying binomials.

DIVIDING BINOMIALS

Although it is possible to divide a binomial by another binomial, the results aren't always useful. In fact, most of the time they are anything but useful. For example:

$$\frac{x+5}{x+3} = \frac{x+5}{x+3}$$

$$\frac{x+8}{y-4} = \frac{x+8}{y-4}$$

On the other hand, a useful result can be obtained sometimes:

$$\frac{x^2 - 16}{x - 4} = x + 4$$

If you can't see why this example is true, try multiplying $(x + 4)(x - 4)$ and see what you get. $x^2 - 16$ is a special case known as **the difference of two squares.** Such a case exists when each of the terms that make up the binomial is a perfect square, and the second one is subtracted from the first. Here are three examples of divisions of perfect squares:

$$\frac{x^2 - 25}{x - 5} = x + 5$$

$$\frac{y^2 - 36}{y + 6} = y - 6$$

$$\frac{z^2 - 49}{z - 7} = z + 7$$

In each case the solution is the same as the **divisor** (the term being divided by) but with an opposite sign. Each term of the solution is the square root of the analogous term of the dividend (the number being divided into).

You get a look at dividing polynomials and binomials in the next chapter when studying factoring. You also get to look at dividing binomials by polynomials there.

Up until this point, you've worked exclusively with equations having only one variable. In the real world, however, this is very often not the case. Suppose Ali has a certain amount of money and Dylan has a certain amount of money, and you don't know what either amount is. Then you are dealing with two different unknown quantities, or, algebraically speaking, with two variables. Call the amount of money Ali has a and the amount of money Dylan has d. Do you see a pattern there? Suppose that together they have \$50. Then you could represent that with the equation $a + d = 50$. Do you think that that's enough information to determine how much money each of them has? It's not, but it's a start. Now what if Ali has \$12 more than Dylan? Algebraically, that's $a = d + 12$. Will that equation let you find the amount each has? The answer is that by itself, it will not, but taken together with the earlier information, it will.

Shown at right is what is known as **a system of equations.** They are also known as **simultaneous equations.** That's because they must be solved together.

$$\begin{cases} a + d = 50 \\ a = d + 12 \end{cases}$$

The brace to the left of the equations is not essential but is there to show you that the equations must be worked together. There are essentially three ways of doing that. You deal with two of those ways in this chapter. The third has to wait until you are practicing your graphing skills (in Chapter 8).

SOLVING SYSTEMS OF EQUATIONS BY ADDITION OR SUBTRACTION

To find the amount of money Ali and Dylan each has, you need to get both variables lined up together on the same side of the equal sign. In this case, that is easily done by subtracting d from both sides of the second equation:

$$a - d = d + 12 - d$$
$$a - d = 12$$

Now look at how the two equations line up. The next step is to add them together. How can you justify doing that? Remember, what is on one side of an equal sign has the same value as what's on the other, and when equals are added to equals, you may recall, the results are equal. (That's a rephrasing of the additive axiom, which you studied in Chapter 5.) So:

$$\begin{array}{r} a + d = 50 \\ a - d = 12 \\ \hline 2a \phantom{{}-d} = 62 \end{array}$$

Now you have an equation with only one variable, so let's solve it by dividing both sides by the same quantity, 2:

$$\frac{2a}{2} = \frac{62}{2}$$
$$a = 31$$

Now that you've found the value of a, substitute that value in either of the original two equations. Here's the first:

$$a + d = 50$$
$$31 + d = 50$$
$$31 - 31 + d = 50 - 31$$
$$d = 19$$

So now you know that Dylan had $19 and Ali had $31. Does Ali have $12 more than Dylan? $31 - 19 = 12$, so she does. Do they have a total of $50 between them? $31 + 19 = 50$, so that's another yes.

Try another one that's just a little bit trickier:

$$2x + 3y = 36$$
$$3x + 2y = 44$$

This time, both variables are on the same side of the equal sign, so there's no work needed there. Wait a second, how are you going to make one of those variables disappear for awhile? Well, here's a thought. Multiply the top equation by 2 and the bottom one by -3. That should help make the x term disappear.

$$3(2x + 3y = 36)$$
$$\underline{-2(3x + 2y = 44)}$$

Remember that everything will change signs when multiplied by the -2.

$$6x + 9y = 108$$
$$\underline{-6x - 4y = -88}$$
$$5y = 20$$

So now you need to divide through by 5.

$$\frac{5y}{5} = \frac{20}{5}$$
$$y = 4$$

Now substitute what you have found into either of the original equations. Here's the first:

$$2x + 3y = 36$$
$$2x + 3(4) = 36$$

Next you multiply and subtract 12 from both sides:

$$2x + 12 = 36$$
$$2x + 12 - 12 = 36 - 12$$
$$2x = 24$$

Finally, divide both sides by the coefficient of x (that's 2).

So, $x = 12$, and $y = 4$.

$$\frac{2x}{2} = \frac{24}{2}$$
$$x = 12$$

Solving Systems of Equations by Substitution

Addition/subtraction is a very handy way to solve certain systems of equations, but it is not always convenient to use. An alternative method of solution is that of substitution.

Check out this pair of equations:

$$x = y + 6$$
$$3x - 2y = 21$$

Solving them by addition would be a bit cumbersome, although it is certainly doable. On the other hand, the first equation *is already solved* for x. It might not be the solution that you'd ultimately like to have, but it is a solution; $x = y + 6$. Take that solution and substitute—plug—it into the second equation:

$$3x - 2y = 21$$
$$3(y + 6) - 2y = 21$$

Next, multiply through by the 3 and combine like terms:

$$3y + 18 - 2y = 21$$
$$3y - 2y + 18 = 21$$
$$y + 18 - 18 = 21 - 18$$
$$y = 3$$

Try one more of those. Look at this set of equations:

$$5n - 6p = 11$$
$$2p = 36 - 4n$$

Sure, you could solve it by addition/subtraction, but you'd have to do some pretty fancy moving around and multiplying. The second equation lends itself to a quick solution, so substitution is the easier process to pursue. To solve the second equation for p, you just need to divide by 2:

$$2p = 36 - 4n$$
$$\frac{2p}{2} = \frac{36 - 4n}{2}$$
$$p = 18 - 2n$$

Now that you've found an expression that is equal to p, you just substitute that expression into the first equation and solve:

$$5n - 6p = 11$$
$$5n - 6(18 - 2n) = 11$$
$$5n - 108 + 12n = 11$$
$$17n - 108 + 108 = 11 + 108$$
$$17n = 119$$
$$\frac{17n}{17} = \frac{119}{17}$$
$$n = 7$$

Now that you've found the value of n, you can substitute it into the streamlined second equation:

$$p = 18 - 2n \qquad p = 18 - 2(7)$$
$$p = 18 - 14$$
$$p = 4$$

That's all there is to it.

Practice Questions

Solve each of the following, and—where appropriate—express your answer in simplest terms.

1 $5a^5b^4 - (-2a^5b^4) = $ ____

2 $5j^3k^2 - 2j^3k^2 = $ ____

3 $4x^4y^7 + 3x^4y^7 = $ ____

4 $5c^3d^5 - 4c^3d^5 = $ ____

5 $(-4e^3f^2)(-2e^3f^5) = $ ____

6 $(-2w^3x^2)^4 = $ ____

7 $8x^5y^6z^7 \div 4x^2y^3z^4 = $ ____

8 $\dfrac{-9r^9s^7t^8}{-6r^3s^4t^5} = $ ____

9 $(6x + 5) + (3x + 8) = $ ____

10 $9v + 11 - (4v - 5) = $ ____

11 $(3y + 4)(5y + 6) = $ ____

12 $(2x - 3)(2x + 3) = $ ____

Solve the following systems of equations by addition/subtraction.

13 $3x + 6y = 30$
$x + 5y = 19$

14 $6a - 3b = 3$
$7a + 3b = 23$

15 $4x - 3y = 13$
$2y + 4x = 38$

Solve the following systems of equations by substitution.

16 $m + 3n = 23$
$4n - 3m = 9$

17 $3x + 6y = 30$
$x + 5y = 19$

18 $6c - d = 47$
$3d + 4c = 57$

Chapter Practice Answers

1 $5a^5b^4 - (-2a^5b^4) = 5a^5b^4 + 2a^5b^4$
$\qquad 5a^5b^4 + 2a^5b^4 = 7a^5b^4$

2 $5j^3k^2 - 2j^3k^2 = 3j^3k^2$

3 $4x^4y^7 + 3x^4y^7 = 7x^4y^7$

4 $5c^3d^5 - 4c^3d^5 = c^3d^5$

5 $(-4e^3f^2)(-2e^3f^5) = (-4)(-2)e^{3+3}f^{2+5}$
$\qquad (-4)(-2)e^{3+3}f^{2+5} = 8e^6f^7$

6 $(-2w^3x^2)^4 = (-2)^4(w^3)^4(x^2)^4$
$\qquad (-2)^4(w^3)^4(x^2)^4 = 16w^{12}x^8$

7 $8x^5y^6z^7 \div 4x^2y^3z^4 = 2x^{5-2}y^{6-3}z^{7-4}$
$\qquad 2x^{5-2}y^{6-3}z^{7-4} = 2x^3y^3z^3$

8 $\dfrac{-9r^9s^7t^8}{-6r^3s^4t^5} = \dfrac{9}{6}r^{9-3}s^{7-4}t^{8-5}$

$\qquad \dfrac{9}{6}r^{9-3}s^{7-4}t^{8-5} = \dfrac{3}{2}r^6s^3t^3$

9 $(6x + 5) + (3x + 8) = 9x + 13$

10 $9v + 11 - (4v - 5) = 9v + 11 - 4v + 5$
$\qquad 9v + 11 - 4v + 5 = 5v + 16$

11 $(3y + 4)(5y + 6) = 15y^2 + 38y + 24$

12 $(2x - 3)(2x + 3) = 4x^2 - 9$

13
$$3x + 6y = 30$$
$$x + 5y = 19$$
$$3x + 6y = 30$$
$$-3(x + 5y = 19)$$
$$3x + 6y = 30$$
$$\underline{-3x - 15y = -57}$$
$$-9y = -27$$
$$y = 3$$
$$x + 5y = 19$$
$$x + 5(3) = 19$$
$$x + 15 = 19$$
$$x + 15 - 15 = 19 - 15$$
$$x = 4$$

14
$$6a - 3b = 3$$
$$\underline{7a + 3b = 23}$$
$$13a = 26$$
$$a = 2$$
$$6a - 3b = 3$$
$$6(2) - 3b = 3$$
$$12 - 3b = 3$$
$$12 - 12 - 3b = 3 - 12$$
$$-3b = -9$$
$$b = 3$$

15

$$4x - 3y = 13$$
$$2y + 4x = 38$$
$$4x - 3y = 13$$
$$-1(4x + 2y = 38)$$
$$4x - 3y = 13$$
$$\underline{-4x - 2y = -38}$$
$$-5y = -25$$
$$y = 5$$
$$4x - 3y = 13$$
$$4x - 3(5) = 13$$
$$4x - 15 = 13$$
$$4x - 15 + 15 = 13 + 15$$
$$4x = 28$$
$$x = 7$$

16

$$m + 3n = 23$$
$$4n - 3m = 9$$
$$m + 3n = 23$$
$$m + 3n - 3n = 23 - 3n$$
$$m = 23 - 3n$$
$$4n - 3m = 9$$
$$4n - 3(23 - 3n) = 9$$
$$4n - 69 + 9n = 9$$
$$13n - 69 + 69 = 9 + 69$$
$$13n = 78$$
$$n = 6$$
$$m + 3n = 23$$
$$m + 3(6) = 23$$
$$m + 18 = 23$$
$$m + 18 - 18 = 23 - 18$$
$$m = 5$$

⑰
$$3x + 6y = 30$$
$$x + 5y = 19$$
$$x + 5y - 5y = 19 - 5y$$
$$x = 19 - 5y$$
$$3x + 6y = 30$$
$$3(19 - 5y) + 6y = 30$$
$$57 - 15y + 6y = 30$$
$$57 - 9y = 30$$
$$57 - 57 - 9y = 30 - 57$$
$$-9y = -27$$
$$y = 3$$
$$3x + 6y = 30$$
$$3x + 6 \cdot 3 = 30$$
$$3x + 18 = 30$$
$$3x + 18 - 18 = 30 - 18$$
$$3x = 12$$
$$x = 4$$

⑱
$$6c - d = 47$$
$$3d + 4c = 57$$
$$6c - d = 47$$
$$6c - 6c - d = 47 - 6c$$
$$-d = 47 - 6c$$
$$d = -47 + 6c = 6c - 47$$
$$3d + 4c = 57$$
$$3(6c - 47) + 4c = 57$$
$$18c - 141 + 4c = 57$$
$$22c - 141 = 57$$
$$22c - 141 + 141 = 57 + 141$$
$$22c = 198$$
$$c = 9$$
$$3d + 4c = 57$$
$$3d + 36 = 57$$
$$3d + 36 - 36 = 57 - 36$$
$$3d = 21$$
$$d = 7$$

chapter 7

Polynomials and Factoring

*P*olynomials are expressions containing *many* terms. This chapter is concerned with all the usual arithmetic operations as they apply to polynomials, and one that might be new to you called *factoring*. Factoring is separating a polynomial (or anything else for that matter) into its component parts. It is, if you will, the undoing of multiplication, but it is definitely different from division, which shares that definition. In this chapter, you deal with factoring out (removing) the greatest common factor, factoring the difference of two squares, and factoring polynomials of the form $ax^2 + bx + c$. You're sure to love it.

You already touched on polynomials in the last chapter. Touching them is one thing; actually seeing them and using them is quite another. Any numeric expression containing more than one term is a polynomial. That makes a binomial, a numeric expression that contains two terms, the simplest form of a polynomial. A trinomial contains exactly three terms, but it's a word that is rarely used. Remember, terms are separated by + or – signs. (As a general practice, any expression with three or more terms is referred to as a polynomial.)

The most common way to write a polynomial is in descending order. That means the power of a term decreases as you move from left to right, for instance:

$$3x^4 + 5x^3 - 4x^2 - 8x + 7$$

When *two* variables are involved, however, it is not unusual to see the polynomial arranged with the first variable in *descending* order and the second in *ascending* order, as in the equation here:

$$2x^5 - 4x^4y + 3x^3y^2 + 2x^2y^3 + 6xy^4 + 3y^5 - 9$$

Notice that in both of the equations the lone constant came last, as is also the convention. When more than two variables are involved, things can become confusing, as you'll see in the section on multiplying polynomials, later in this chapter.

Adding and Subtracting Polynomials

To add or subtract polynomials, arrange them in columns with like terms above one another, just like you did with binomials, except you have more terms to worry about. Two examples are shown at right.

$$
\begin{array}{ll}
2m^2 + \ mn + 3n^2 & 3b^2 - 4bc + 5c^2 \\
3m^2 + \ mn + 5n^2 & 4b^2 + 2bc - 3c^2 \\
\hline
5m^2 + 2mn + 8n^2 & 7b^2 - 2bc + 2c^2
\end{array}
$$

In the left addition, there are nothing but plus signs, so each vertical sum is larger than either addend. In the right equation there are two minus signs, so the laws of signed number addition apply.

To the right is an example of subtraction of polynomials. In order to subtract, you first distribute the minus sign. (It's as if you're multiplying everything inside the parentheses by −1.) Then you treat the resulting vertical pairings as if they were signed number additions.

$$\begin{array}{l} 5x^2 - 4xy + 8y^2 - 9 \\ -\left(4x^2 + 5xy + 3y^2 - 7\right) \end{array} \longrightarrow \begin{array}{l} 5x^2 \;\; -4xy \;\; +8y^2 \;\; -9 \\ -4x^2 - 5xy - 3y^2 +7 \\ \hline x^2 - 9xy + 5y^2 - 2 \end{array}$$

It's not always necessary to rearrange terms in an addition or subtraction of polynomials. Sometimes the like pairs can be paired up in your head, and the results recorded.

$$3m^2n^3 + 4m^3n^2 + 7 + 2m^2n^3 + 3m^3n^2 + 5 = 5m^2n^3 + 7m^3n^2 + 12$$

Multiplying Polynomials

To multiply polynomials, multiply every term in one polynomial by each term in the other polynomial. Then simplify, if possible, by combining like terms. Of course, if you're dealing with a binomial times a binomial, you can use the **FOIL** method that you studied in the previous chapter.

$$\begin{array}{r} 3x^3 - 2x^2 + 7x + 4 \\ \times \qquad\qquad 5x + 3 \\ \hline 9x^3 - 6x^2 + 21x + 12 \\ 15x^4 - 10x^3 + 35x^2 + 20x \qquad\quad \\ \hline 15x^4 - x^3 + 29x^2 + 41x + 12 \end{array}$$

Here's an example of a polynomial multiplied by a binomial. First multiply each member of the polynomial by 3. Then multiply each of its terms by $5x$. Make sure to line up partial products on the second line of the solution with the terms on the first line to which they'll be added. Finally, add like terms.

I would guess that you get the picture from that, but let's work out one more. Again, start at the right and multiply each of the top terms by $-3c$:

$$
\begin{array}{r}
a + 2b - 3c - 4 \\
\times \quad a + 2b - 3c \\
\hline
-3ac - 6bc + 9c^2 + 12c
\end{array}
$$

Next, multiply each of the top terms by $+2b$:

$$
\begin{array}{r}
a + 2b - 3c - 4 \\
\times \quad a + 2b - 3c \\
\hline
-3ac \qquad - 6bc \qquad + 9c^2 + 12c \\
+2ab + 4b^2 - 6bc - 8b
\end{array}
$$

Starting to look a little strange, isn't it? Yet so far only one term is alike in both partial product lines. Next, multiply each of the top terms by a, and then add:

$$
\begin{array}{r}
a +2b -3c -4 \\
\times \qquad a +2b -3c \\
\hline
-3ac \qquad -6bc \qquad +9c^2 +12c \\
+2ab +4b^2 -6bc -8b \\
-3ac +2ab \qquad\qquad\qquad\qquad +a^2 -4a \\
\hline
-6ac +4ab +4b^2 -12bc -8b +9c^2 +12c +a^2 -4a
\end{array}
$$

That can be rearranged as you see at right. The only reason for the rearrangement is to try to bring a bit of organization to the polynomial.

$$a^2 - 4a + 4ab - 6ac - 8b - 12bc + 4b^2 + 12c + 9c^2$$

Dividing Polynomials

Polynomials can be divided by monomials and by polynomials. Essentially two very different processes can be done in two very different ways. Here, you take a look at the simpler of those two processes first.

Polynomials Investigated (continued)

DIVIDING POLYNOMIALS BY MONOMIALS

To divide a polynomial by a monomial, simply divide each of the terms in the polynomial by the monomial, for instance:

$$\left(9y^2 - 3y\right) \div 3y = \frac{9y^2 - 3y}{3y}$$

$$= \frac{9y^2}{3y} - \frac{3y}{3y}$$

$$= 3y - 1$$

Note that for a polynomial to be *divisible* by a monomial, each term of the polynomial must be divisible by that monomial. Try this one to the right before you look at its solution (below right).

$$\left(24x^8 + 16x^6 + 8x^4\right) \div 4x^3 =$$

Did you solve it? You could solve it in your head by dividing each numerical coefficient by 4 and subtracting 3 from each exponent, or you could do what is shown at right.

$$\left(24x^8 + 16x^6 + 8x^4\right) \div 4x^3 = \frac{24x^8 + 16x^6 + 8x^4}{4x^3}$$

$$= \frac{24x^8}{4x^3} + \frac{16x^6}{4x^3} + \frac{8x^4}{4x^3}$$

$$= 6x^5 + 4x^3 + 2x$$

Is this what you got? I hope so.

DIVIDING POLYNOMIALS BY POLYNOMIALS

In order to divide a polynomial by a polynomial, both must have their variable terms arranged in *descending* order, with constants last. Since this is a rather esoteric process, use a binomial divisor.

$$y + 3 \overline{\smash{)}3y^3 + 11y^2 + 11y + 15}$$

I trust you recognize the long division bracket, and that's exactly how this process is done. It's the repetitive practice of divide by the first term, write the partial quotient above the first term, multiply the divisor by that partial quotient, write the product, subtract, bring down. Does that bring back bad memories? It does for me, too.

Start by dividing y into $3y^3$ and multiplying.

$$y+3\overline{)3y^3+11y^2+11y+15} \\ \underline{3y^3+9y^2}$$

quotient: $3y^2$

$3y^2$ times $y + 3 = 3y^3 + 9y$, which you write below the dividend. Next, subtract and bring down.

$$\begin{array}{r} 3y^2 \\ y+3\overline{)3y^3+11y^2+11y+15} \\ \underline{-(3y^3+9y^2)}\quad\downarrow\quad\downarrow \\ 2y^2+11y+15 \end{array}$$

How many times does y go into $2y^2$? $2y$ you say? So do I; then multiply, subtract, etc.

$$\begin{array}{r} 3y^2+2y+5 \\ y+3\overline{)3y^3+11y^2+11y+15} \\ \underline{-(3y^3+9y^2)}\quad\downarrow\quad\downarrow \\ 2y^2+11y+15 \\ \underline{-(2y^2+6y)}\quad\downarrow \\ 5y+15 \\ \underline{-(5y+15)} \end{array}$$

That "etc." was a second bring down, and a third divide, multiply, subtract; $5y$ divided by y is 5, which completed the quotient. Then multiply that 5 times the divisor, write it at the very bottom, and subtract to get a remainder of 0. Does it look strange to you that the variables in the quotient don't line up with those in the dividend? Feel free to move them if it'll make you more comfortable, but it's unnecessary.

Here's another division with a new twist:

$$c+1\overline{)c^3-c}$$

In case you didn't notice, the dividend's terms jump straight from c to the third power to c to the first power. That is not capable of being divided *in that form*. There may not be any stages missing from the dividend, so what are you to do? The answer is to put in a second power term that has no real value, as shown here:

$$c+1\overline{\smash{)}c^3+0c^2-c}$$

Now you can go ahead and perform the subtraction. (Note the absence of a constant is not a problem.) How many cs are there in c^3?

$$\begin{array}{r} c^2 \\ c+1\overline{\smash{)}c^3+0c^2-c} \end{array}$$

Next, multiply the partial quotient times the divisor, and write the product below the appropriate members of the dividend.

$$\begin{array}{r} c^2 \\ c+1\overline{\smash{)}c^3+0c^2-c} \\ c^3+c^2 \end{array}$$

Now, subtract and bring down the next term(s).

$$\begin{array}{r} c^2 \\ c+1\overline{\smash{)}c^3+0c^2-c} \\ -\left(c^3+c^2\right)\downarrow \\ \hline -c^2-c \end{array}$$

The result of the subtraction is the new partial dividend. Next, ask yourself how many times c goes into $-c^2$. The answer to that is $-c$. Don't take my word for it. Multiply c by $-c$ and see what you get (pun intended). So write that amount in the quotient and multiply it by the divisor to get:

$$\begin{array}{r} c^2-c \\ c+1\overline{\smash{)}c^3+0c^2-c} \\ -\left(c^3+c^2\right)\downarrow \\ \hline -c^2-c \\ -c^2-c \end{array}$$

Finally, subtract and get a remainder of 0:

$$\begin{array}{r} c^2 - c \\ c+1\overline{)c^3 + 0c^2 - c} \\ -\underline{(c^3 + c^2)} \quad \downarrow \\ -c^2 \ - c \\ -\underline{(-c^2 \ - c)} \\ 0 \end{array}$$

Here's one last one:

$$1+h\overline{)1 + 2h + h^2}$$

Is there a problem here? You're darned right there is. Both the divisor and the dividend are listed *in ascending order* of the powers of h. That will not do, but the solution is quite simple. Just turn them both around and divide in the usual way. The steps have been worked out here.

$$\begin{array}{r} h+1 \\ h+1\overline{)h^2 + 2h + 1} \\ -\underline{(h^2 + h)} \quad \downarrow \\ h \ + 1 \\ -\underline{(h \ + 1)} \\ 0 \end{array}$$

Note that it is possible to divide polynomials and end up with a remainder. In such a case, the remainder would be written as the last term with its sign in front of it. Suppose, for example, that the last division had left you with a remainder of –5. Then the quotient would have been written as this:

$$h + 1 - \frac{5}{h+1}$$

Factoring

Factoring means finding two or more expressions that, when multiplied together, make up the initial polynomial. That might sound to you at first like division, and in its simplest form, it is. Then it gets somewhat more complicated than that. Remember, division results in resolving a polynomial into exactly *two factors*. **Oops, there's that word again. Factoring results in** *two or more factors*. **That is a difference!**

Factoring Out the Greatest Common Factor

You may have noticed that *factor* may be used as a noun or as a verb, and it is used both ways in the heading of this section. As a noun, a **factor** is one of two or more monomials that are multiplied together to form some other expression. To factor, the verb, is to break apart an expression into its multiplication components.

Here is an example of factoring by removing a common factor:

$$8x^3y^2 + 4x^4y^3 = 2xy(4x^2y + 2x^3y^2)$$

In this equation, the expression $8x^3y^2 + 4x^4y^3$ has been factored by removing the quantity $2xy$. You say $2xy$ was *factored out* of the original expression. Two questions immediately come to mind. First, was it factored correctly? Second, has it been factored completely? The answer to the first question is yes, and the answer to the second, no. It was factored correctly because when you *distribute* $2xy$ over $4x^2y + 2x^3y^2$ you get $8x^3y^2 + 4x^4y^2$.

There is, however, still more that can be *factored out*, as you see here:

$$8x^3y^2 + 4x^4y^3 = 4x^3y^2(2y + x)$$

You call $4x^3y^2$ the **greatest common factor**, or **GCF**. You may remember that expression as having been used before, when you learned to work with fractions. This is the same idea. A monomial or a polynomial is factored completely when the greatest possible monomial has been factored out. When that monomial happens to be identical to one of the terms in the expression being factored, then 1 is used as a placeholder, as in the factorization to the right:

$$14ab^2 + 21a^2b - 7ab = 7ab(2b + 3a - 1)$$

To make certain that this factorization is correct, *distribute 7ab* over (*multiply it times*) each term in parentheses.

Factoring the Difference of Two Squares

You've dealt with the difference of two squares before when you looked at **FOIL**, in case you don't recall. It was when you multiplied two binomials together and got a binomial because what should have been the middle term dropped out.

Following is an example of the difference of two squares and how to factor it:

$$x^2 - 49 = (x + 7)(x - 7)$$

Squares and square roots are discussed in Chapter 1, just in case you need to review the topic. x^2 is a perfect square ($x \cdot x$), and so is 49 ($7 \cdot 7$). Because a minus sign is between them (**subtraction** means *finding the difference*), the expression $x^2 - 49$ is known as the **difference of two squares.** To factor the difference of two squares, multiply the sum of their square roots times the difference of their square roots.

Try factoring these three binomials:

$$y^2 - 64 =$$
$$n^2 - 100 =$$
$$x^2 + 81 =$$

Did you solve them? All three? Did you pick up on my little nasty trick? If you did—good for you. If you didn't, tsk, tsk, tsk!

$$y^2 - 64 = (y + 8)(y - 8)$$
$$n^2 - 100 = (n + 10)(n - 10)$$
$$x^2 + 81 = \text{Are you kidding?}$$

$x^2 + 81$ is *not* the difference of two squares. It's the sum of two squares, and as such, is not solvable. Didn't you notice the color difference?

Factoring
(continued)

Try one more:

$$9x^2 - 121 =$$

Did the 9 throw you? It shouldn't have. It's also a perfect square, so:

$$9x^2 - 121 = (3x + 11)(3x - 11)$$

Factoring Polynomials of the Form $ax^2 + bx + c$

To factor three-term polynomials of the form $ax^2 + bx + c$, the first thing to do is to see whether there's a monomial that can be factored out. If there is, factor it out as you did earlier in "Factoring Out the Greatest Common Factor." Then, if you're lucky enough to be left with a first term that begins with an x^2, set up double parentheses and factor the first term, as you see here:

$(x\ )(x\ )$

The next thing to do is to determine the signs that will follow the two variables. Look at the sign before the c term of the polynomial. If it's a minus, then one sign will be a $+$ and the other a $-$.

If $-c$ then $(x +\ )(x -\ )$

If the sign before the c term is plus, then there are two possibilities:

If $+c$ then $(x +\ )\ (x +\ )$ or $(x -\ )\ (x -\ )$

To know which it is, look at the b term. A $+$ before it means two pluses; a $-$ sign before it means they're both minuses.

If $+c$ and $+b$ then $(x +\ )(x +\ )$

If $+c$ and $-b$ then $(x -\ )(x -\ )$

Finally, break the last term into two factors. If both signs are the same, the two factors must add up to make the numerical coefficient of the b term; if the signs are different, they must subtract to make the numerical coefficient of the b term. Here are some examples:

$$y^2 + 6y - 7$$
$$(y +\ )(y -\ )$$
$$(y + 7)(y - 1)$$

Notice the $-c$ term.

$$3x^2 + 6x + 3$$
$$3(x^2 + 2x + 1)$$
$$3(x +)(x +)$$
$$3(x + 1)(x + 1)$$

This time you were able to factor out the ax^2 term's numerical coefficient.

$$z^2 - 7z + 12$$
$$(z -)(z -)$$
$$(z - 3)(z - 4)$$

How could you tell that the factors of the c term were 3 and 4 instead of 12 and 1 or 6 and 2? The answer is, they had to add up to the b term, 7. And last but not least:

$$2x^2 - 9x - 5$$
$$(x +)(x -)$$
$$(2x + 1)(x - 5)$$

That kind of threw in a new wrinkle, but remember **FOIL** (First Outer Inner Last). The outer terms are $2x$ and -5, which multiply to make $-10x$; when added to the product of the inners, x, the sum is $-9x$, the b term.

Arrange the terms of the polynomials in descending order of the main variable and ascending order of other variables when appropriate.

1 $7f^2 + 6f^3 - 9 - 8f - 3f^4 + 5f^5$

2 $7 - 2m^4 - 9m + 7m^2 + 8m^5 + 6m^3 + 4m^6$

3 $3b^2y^7 + 4y^{10} - 13by^4 - 5ay^{12} - 6a^3y^8 + 4y^9 + 3y^3$

Perform the indicated operations on the following polynomials.

4 Add $6x^4 + 8x^2 - 11 - 4x^3 - 3x^2$ and $6x^3 - 4x - 11 + 2x^2 - 3x^4$

5 Subtract $-9y^3 + 4y + 3y^4 - 7 + 5y^2$ from $6y^3 - 8y + 11 - 4y^4 + 3y^2$

6 Add $3x^2 - 8x^5 - 7 + 11x + 4x^3$ and $3x^4 - 5x + 9x^2 - 8x^3 + 2x^5$

Solve each of the following as indicated.

7 $(z + 3)(z - 3) = $ ___

8 $(x - 9)(x - 3) = $ ___

9 $(y - 5)(y - 6) = $ ___

10 $(p^2 + 4)(p - 2)(p + 2) = $ ___

11 $x - 5 \overline{)2x^2 - 7x - 15}$

12 $y + 2 \overline{)y^2 + 9y + 14}$

Factor each of the following as completely as possible.

13 $4m^2n^4p^6 = $ ___

14 $6x^3y^5z^7 = $ ___

15 $5b^4c^7d^7 - 7b^5c^3d^9 = $ ___

16 $16r^2 - 81 = $ ___

17 $m^2 + 12m + 35 = $ ___

18 $k^2 + 6k - 27 = $ ___

19 $t^2 - 7t + 12 = $ ___

20 $49s^2 - 121 = $ ___

21 $x^2 - 5x - 24 = $ ___

22 $2y^2 + 13y + 15 = $ ___

① $5f^5 - 3f^4 + 6f^3 + 7f^2 - 8f - 9$

② $4m^6 + 8m^5 - 2m^4 + 6m^3 + 7m^2 - 9m + 7$

③ $-5ay^{12} + 4y^{10} + 4y^9 - 6a^3y^8 + 3b^2y^7 - 13by^4 + 3y^3$

④ First combine $8x^2 - 3x^2$ to get $5x^2$. Then:

$$\begin{array}{r} 6x^4 - 4x^3 + 5x^2 + 0x - 11 \\ -3x^4 + 6x^3 + 2x^2 - 4x - 11 \\ \hline 3x^4 + 2x^3 + 7x^2 - 4x - 22 \end{array}$$

⑤ $-4y^4 + 6y^3 + 3y^2 - 8y + 11 - (3y^4 - 9y^3 + 5y^2 + 4y - 7)$

Next, change all signs on the bottom:

$$-4y^4 + 6y^3 + 3y^2 - 8y + 11 - (3y^4 + 9y^3 - 5y^2 - 4y + 7)$$

Finally, add:

$$\begin{array}{r} -4y^4 + 6y^3 + 3y^2 - 8y + 11 \\ -3y^4 + 9y^3 - 5y^2 - 4y + 7 \\ \hline -7y^4 + 15y^3 - 2y^2 - 12y + 18 \end{array}$$

⑥ This time, pair up like terms and then combine. Using different colors helps to keep track of what came from where:

$$3x^2 - 8x^5 - 7 + 11x + 4x^3 \text{ and } 3x^4 - 5x + 9x^2 - 8x^3 + 2x^5$$
$$2x^5 - 8x^5 + 3x^4 - 8x^3 + 4x^3 + 3x^2 + 9x^2 + 11x - 5x - 7$$

$$-6x^5 + 3x^4 - 4x^3 + 12x^2 + 6x - 7$$

Chapter Practice
(continued)

7 You might use the **FOIL** method on this multiplication, or you might recognize it as the result of factoring the difference of two squares, in which case the solution should jump into your head. First, factor the difference of two squares:

$$(z + 3)(z - 3) = z^2 - 9$$

Now, use the **FOIL** approach:

$$(z + 3)(z - 3) = z^2 - 3z + 3z - 9 = z^2 - 9$$

8 **FOIL** is the way to go:

$$(x - 9)(x - 3) = x^2 - 3x - 9x + 27 = x^2 - 12x + 27$$

9 And again:

$$(y - 5)(y - 6) = y^2 - 6y - 5y + 30 = y^2 - 11y + 30$$

10 There are three ways to solve this multiplication, but the easiest is to recognize that the second and third binomials multiply together to form the difference of two squares, and by multiplying them together you get another difference of two squares, so:

$$(p^2 + 4)(p - 2)(p + 2) = (p^2 + 4)(p^2 - 4) = p^4 - 16$$

11
$$\begin{array}{r} 2x + 3 \\ x - 5 \overline{)\, 2x^2 - 7x - 15} \\ -(2x^2 - 10x) \ \downarrow \\ \hline 3x - 15 \\ -(3x - 15) \\ \hline 0 \end{array}$$

12
$$\begin{array}{r} y + 7 \\ y + 2 \overline{)\, y^2 + 9y + 14} \\ -(y^2 + 2y) \ \downarrow \\ \hline 7y + 14 \\ -(7y + 14) \\ \hline 0 \end{array}$$

136

⑬ $4m^2n^4p^6 = (2mn^2p^3)^2$

⑭ You have many possible solutions here, as long as the numerical parts multiply to 6, and the exponents add up to the numbers 3, 5, and 7, respectively.

$$6x^3y^5z^7 = (2x^3y^2z^5)(3y^3z^2) \text{ or } (6xy^5z^5)(x^2z^2) \text{ to name two}$$

⑮ $5b^4c^7d^7 - 7b^5c^3d^9 = b^4c^3d^7(5c^4 - 7bd^2)$

⑯ $16r^2 - 81 = (4r + 9)(4r - 9)$

⑰ $m^2 + 12m + 35 = (m + 5)(m + 7)$

⑱ $k^2 + 6k - 27 = (k + 9)(k - 3)$

⑲ $t^2 - 7t + 12 = (t - 4)(t - 3)$

⑳ $49s^2 - 121 = (7s - 11)(7s + 11)$

㉑ $x^2 - 5x - 24 = (x - 8)(x + 3)$

㉒ $2y^2 + 13y + 15 = (2y + 3)(y + 5)$

chapter 8

Cartesian Coordinates

It's about time for a chapter devoted to some of the work of Rene Descartes, and the coordinate system that bears his name, so this is it. Beginning with a philosophical word or two, this chapter goes on to examine the system of naming points on coordinate axes to the graphing of linear equations by substitution, slope and intercept, and finally to the graphing of systems of equations such as those studied in Chapter 6. Along the way, I deal with finding the y-intercept and slope from the equation of a line, finding the equation of a line from its graph, and writing equations in point-slope form. This chapter should certainly assist you in acquiring a brand-new slant on life.

Rene Descartes (pronounced day-CART), March 31, 1596–February 11, 1650, was a French mathematician, philosopher, scientist, and writer. As a philosopher, he is celebrated for his proof of existence "Cogito ergo sum," which means "I think, therefore I am." As a mathematician, he is celebrated for the invention of analytic geometry and Cartesian coordinates (pronounced car-TEE-zhian). Descartes has alternately been referred to as the *Father of Modern Philosophy* or the *Father of Modern Mathematics*. It is his work in the latter area to which this chapter is devoted.

Score Four

Because you exist, it stands to reason that by this point in your life, you have come across the rather simple game known as tic-tac-toe. It is also reasonable to believe that by now, you are aware of the fact that any game of tic-tac-toe played between two players of moderate skill should always result in a draw. Since the game is played on a rather small grid 3 wide by 3 long, getting 3 of your mark in a row is impossible unless you are playing a very careless or thoughtless person. Expanding the playing field and requiring four marks in a row to win the game—now known as Score Four—is much more challenging. In fact, Milton Bradley has turned it into a game in which red and black checkers are used as markers and dropped alternately into a 42-slot board to *Connect Four*.

Playing this game requires nothing more than a sheet of graph paper as a playing field, with lines ruled both horizontally and vertically at regular intervals. The heavy ruled lines are the **axes** (AX-eez) with horizontal blue and vertical red. Unlike tic-tac-toe, which uses X's and O's, you can use X's and dots.

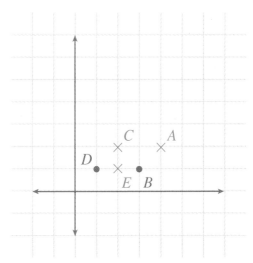

The location of each of the marks on the playing board can be given by a description of where that mark is located. In order to identify where a mark is located on the board, count spaces across and spaces up. The X located at *A*, for example, can be described this way: (4, 2). Start your counting where the horizontal and the vertical axes cross.

Since that's where counting begins, that point is referred to as the **origin**. Place a fingernail or the tip of a ballpoint pen at the origin and count lines as you move to the right, until your pointer is below the X at A. Your pointer will have moved horizontally to 4. Now move your pointer up to the X at A. You will have moved it vertically to 2; hence, the shorthand (4, 2). The expression (4, 2) represents the coordinates of the X at A. That was also the first move made in this game of Score Four.

The second through fifth moves are indicated by the other uppercase letters on the diagram in alphabetical order. Here, we identify the coordinates of each of those points:

$$\bullet B(3,1); \times C(2,2); \bullet D(1,1); \times E(2,1)$$

It's now dot's turn. Remember that the objective of the game is to get four of your marks in a row, horizontally, vertically, or **diagonally**. Where does dot have to go to prevent losing the game? Take your time and think about it. When you have figured out where dot needs to go, you may look at the next diagram.

If a dot were not placed at either F or G, then the other player would be able to place an X, giving an unbeatable open-ended XXX.

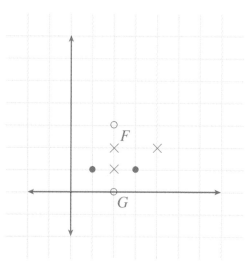

What are the coordinates of points F and G? Place a dot at point G, by calling out coordinates (2, 0) [the other option would have been (2, 3)]. But the dot player isn't out of trouble yet. Where should the next X go—if the X player is on the ball?

H marks the spot and completes the trap. By calling out the coordinates (3, 2), the X player has sprung the dreaded open-ended, three-in-a-row trap.

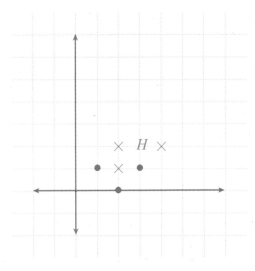

The opposing player can put her dot at location *I* or *J*. Whichever she selects, X will take the other one for the win.

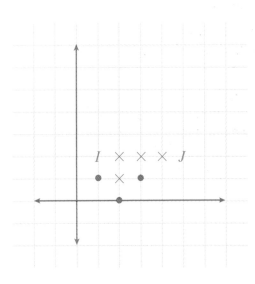

What's especially neat about Score Four is that the playing field's arrangement can be altered each time, just by shifting the locations of the axes. See the right column for the coordinates of *I* and *J*.

I(1,2); *J*(5,2)

Look at the figure at right. All extraneous marks are removed, just leaving the 4 winning dots. What are their coordinates? Reading from left to right, the coordinates are (–1, –1), (0, 0), (1, 1), and (2, 2).

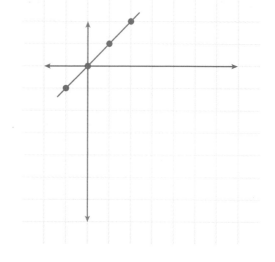

Now look at this figure. All extraneous marks are removed, just leaving the 4 winning ×s. What are their coordinates? Reading from left to right, the coordinates are (–2, –1), (–1, –2), (0, –3), and (1, –4).

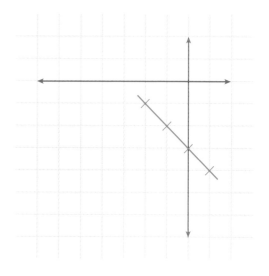

Copy the grids in the two figures to the right of this paragraph and try playing a few games of Score Four with a classmate, parent, or friend. Remember, you must name the coordinates of the point where you wish to place your mark; no pointing. Before any marks are made, you and your opponent must agree that the point you named is where your mark is being made. After all, the purpose of this game is to learn Cartesian coordinates (surprise!).

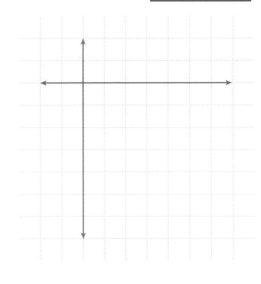

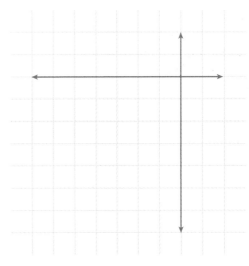

The Coordinate Plane

A **plane** is a flat surface that continues forever in two directions and exists in theory only. In the real universe, a plane isn't possible, but for your purposes, that doesn't matter. The two directions in which the plane extends are horizontal and vertical, which are named, respectively, x and y. Does this state of the plane seem a bit familiar to you? It certainly should. It's the playing field for Score Four, but it has been extended infinitely in all directions. Now names are attached to the axes: the x-axis and y-axis, respectively. Now each and every point on the coordinate plane can be named by a pair of coordinates, namely the x-coordinate and the y-coordinate, in the form (x, y).

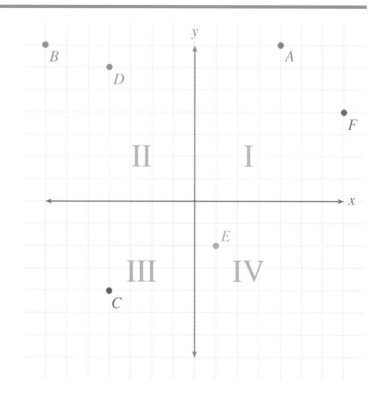

Think of the coordinate plane as being separated into four **quadrants**—a word meaning, of all things, fourths. These quadrants are marked off in the following figure with the Roman numerals, starting with the I in **standard position** and moving counterclockwise.

Notice that the signs of the (x, y) coordinates are quite predictable, as per the scheme shown here:

$I(x,y)$; $II(-x,y)$; $III(-x,-y)$; $IV(x,-y)$

Keeping that color scheme in mind, identify each of the lettered dots in the figure by writing the ordered pair of its coordinates, in the form (x, y). You'll find the correct answers at the bottom of this page.*

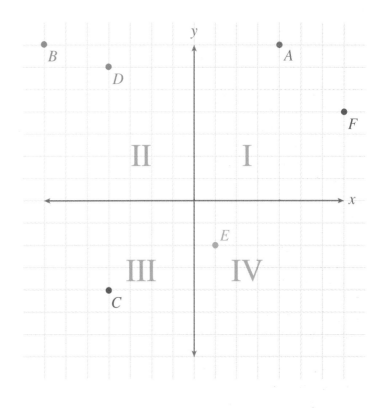

Graphing Linear Equations by Substitution

Although not 100% essential, graph paper will come in handy for use during the rest of this chapter.

Consider this equation:
$$y = 2x + 3$$

*•$A(4,7)$; •$B(-7,7)$; •$C(-4,-4)$; •$D(-4,6)$; •$E(1,-2)$; •$F(7,4)$

In this equation (and most others), *x* is known as the independent variable, while *y* is the dependent one. What this means is you can substitute anything you want for *x* and then find the value of *y* at that location. Substitute 2 for *x*, and see what *y* is. Substitute 0 and –2 for *x* to get the corresponding values of *y*, as shown in this figure.

x	y
2	7
0	3
-2	-1

Since you have, in fact, found the Cartesian coordinates of 3 points on the (*x*, *y*) plane, plot them on that plane, as shown here:

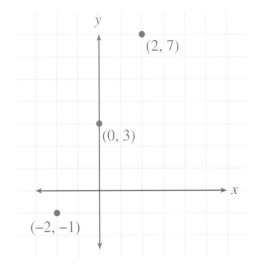

Next, take a ruler and draw the line that is determined by those three points. Note that while it takes only 2 points to determine a line, having 3 ensures that the other two are correct. (If one of the points had not lined up with the other two, you would have known that something was wrong.)

Notice that the **graph** (line) of the equation is infinite, as is represented by the arrowhead at either end.

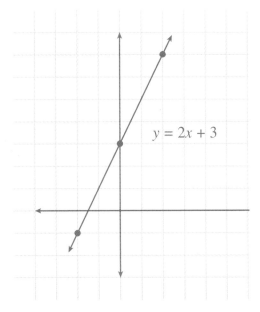

TIP

You've probably noticed how the use of *x* as the variable's name is favored when writing equations. There are a couple of reasons for that, but the most compelling is the ability to (and in some cases the need to) represent an equation and its solutions on a coordinate plane.

The graph of any straight-line equation has certain characteristics. Some have a y–intercept (the point at which it crosses the y-axis), some have an x–intercept (the point at which it crosses the x-axis), and some have one of each.

The line to the right has an x–intercept of $(3, 0)$ and a y–intercept of $(0, -2)$. Note that each intercept has the coordinate not the same as its name equal to 0.

Every linear equation has a slope. **Slope** is alternately defined as **rise over run**, or the difference in ys over the difference in xs, or by the expression $\frac{\Delta y}{\Delta x}$ where Δ is the Greek letter **delta**, which stands for "difference."

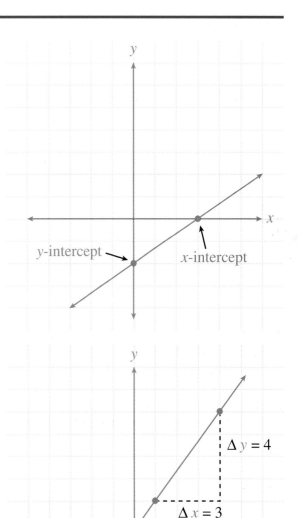

Note that a line with a positive slope runs from lower left to upper right. In the figure to the right, the dotted segments show a rise of 4 for a run of 3. That's a slope of $\frac{4}{3}$. The measurements could have been made between any consecutive or nonconsecutive points on the line, as long as the differences between the coordinates of the two points were clearly discernible; that is: $\frac{\Delta(y_2 - y_1)}{\Delta(x_2 - x_1)}$.

Finding the Slope and Intercepts *(continued)*

This figure at right depicts the graph of a line with a negative slope. You can remember that a line that runs from upper left to lower right has a negative slope, or you can measure rise over run. For a marked run of 4, there's a rise of –2. That's $\frac{\Delta y}{\Delta x} = \frac{-2}{4}$ or $-\frac{1}{2}$.

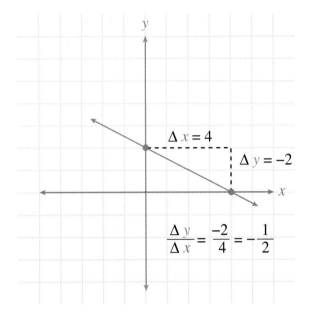

Here is a figure that depicts two lines, l and m, which are parallel to the y–axis and x–axis, respectively. Can you figure out the slopes of those two lines?

The equation of each line is shown in the figure. First look at line m. Consider the differences of any two coordinate pairs.

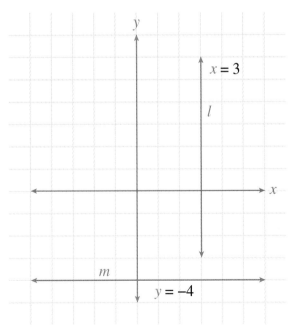

Since all y coordinates are the same, the difference between any two of them is 0, and 0 divided by anything is 0. 0, therefore, is the slope of *any* horizontal line.

$$\frac{\Delta(y_2 - y_1)}{\Delta(x_2 - x_1)} = \frac{-4 - (-4)}{\Delta(x_2 - x_1)} = \frac{0}{\Delta(x_2 - x_1)} = 0$$

Next check out line l. Again, use any pairs of coordinates.

$$\frac{\Delta(y_2 - y_1)}{\Delta(x_2 - x_1)} = \frac{\Delta(y_2 - y_1)}{3 - 3} = \frac{\Delta(y_2 - y_1)}{0}$$

Depending on whom you ask and when you ask them, anything divided by 0 is considered to be either infinite or undefined. At the time of this writing, undefined seems to be the preferred answer, and applies to the slope of any vertical line.

Graphing Linear Equations by Slope and Intercept

Now that you know what slope and intercepts are, it's time to put that knowledge to good use.

An equation is said to be in slope and y–intercept form when it is written in the form $y = mx + b$. The color coding should help you to know what each item is, but this figure spells it all out explicitly.

$$y = mx + b$$

slope   y-intercept

Graph the equation $2y - 6 = 3x$. In order to graph it, you need to get the equation into standard form. First add 6 to each side of the equation:

$$\begin{array}{r} 2y - 6 = 3x \\ +6 \qquad +6 \\ \hline 2y \quad = 3x + 6 \end{array}$$

Next, divide both sides of the equation by 2 in order for y to stand alone.

$$\frac{2y}{2} = \frac{3x + 6}{2}$$
$$y = \frac{3}{2}x + 3$$

To graph the equation, first mark the y–intercept, which is at $(0, 3)$; remember, the b term is the y–coordinate. The x–coordinate is automatically a 0, since it's a y–intercept. When the x-coordinate is placed, use it as a starting point to count 3 up and 2 over; place a point there. Then count 3 up and 2 over; place a point there.

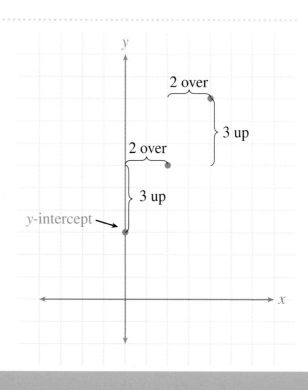

Finally, using a straight-edge ruler, draw the line that passes through those points.

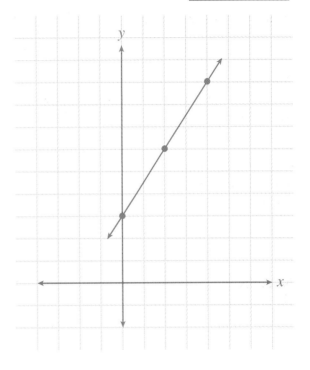

Now that you've seen how that works, try one more. How about $6x = 8 - 2y$? To get that into the $y = mx + b$ form, the easiest thing to do is to first subtract 8 from both sides:

$$6x = 8 - 2y$$
$$\underline{-8 \quad -8}$$
$$6x - 8 = \quad -2y$$

Next, divide both sides by –2 in order to get the y by itself and positive:

$$\frac{6x - 8}{-2} = \frac{-2y}{-2}$$
$$-3x + 4 = y$$

Finally, pivot the equation around the = sign to make it look like it's supposed to:

$$y = -3x + 4$$

Now it's time to mark the *y*–intercept with a dot. Note that the first dot goes at (0, 4). Notice also that you counted one to the right and three down for the next point. Could you have counted one left and three up? Absolutely. The important thing to remember is that since your slope is negative, it must move from upper left to lower right—the *opposite* of a positive slope.

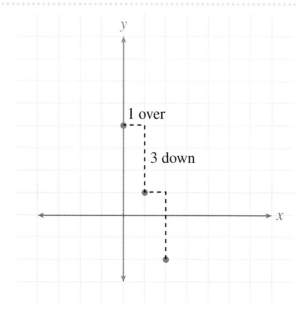

Finally, use a ruler to connect the dots and extend the line.

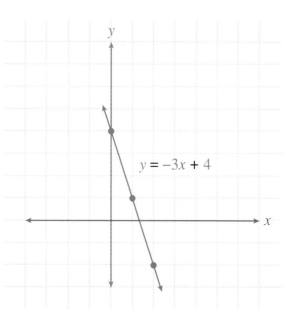

Suppose that you wanted an equation of a line that is parallel to the one you just drew. Which of the following equations would fill the bill?

❶ $12x + 4y = +20$

❷ $2y = -3x + 4$

❸ $3y - 12 = -9x$

❹ $6y = 2x + 36$

Equation #3 is the identical line, as you can see if you add 12 to both sides and divide by 3. If, however, you subtract $12x$ from both sides of equation #1 and divide by 4, you'll find what is shown at right.

How do you know that the graph of this equation is parallel to the one in the preceding graph? It's simple. Both equations have the same slope. Lines with equal or the same slopes are, by definition, parallel.

$$\begin{aligned} 12x + 4y = & \quad +20 \\ -12x \quad\quad & -12x \\ \hline 4y = & -12x + 20 \\ \frac{4y}{4} = & \frac{-12x + 20}{4} \\ y = & -3x + 5 \end{aligned}$$

You also might find it interesting—as well as helpful—to know that perpendicular lines have slopes that are *negative reciprocals* of each other. Which of the four preceding equations is of a line perpendicular to both parallel lines? Well, you can eliminate equations #1 and #3, since you've already discussed them. What is the negative reciprocal of -3? The answer is at right, so you're looking for a line with slope $\frac{1}{3}$. Divide through by 6, and you'll find that equation #4 fits the bill.

$$-\left(\frac{1}{-3}\right) = \frac{1}{3}$$

Finding the Equation of a Line

The figure at right shows the graph of line l. By getting certain information from that graph, you can determine the equation of line l. Not surprisingly, the two pieces of information needed are the slope and the y-intercept of line l. What is the y-intercept of line l? Going to the origin and counting down from it, you find the y-intercept is $(0,-3)$. To find the slope, check the graph of l for rise over run. The slope is 1.

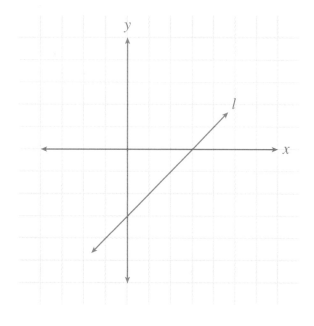

Next you need to substitute the information you just collected into the standard equation for slope and y-intercept and, there, you have it.

$$y = mx + b$$
$$y = (1)x + (-3)$$
$$y = x - 3$$

Still a little nervous? Well, just in case, do one more. Look at this figure, which shows the graph of line m. It's somewhat different from line l, but the method for finding its equation is to apply the same procedure as you did for the previous figure's line l (see the beginning of this section).

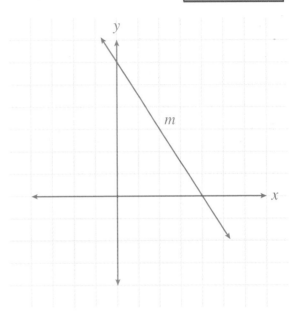

Again, the data needed are the slope and the y–intercept of line m. What is the y–intercept of line m? Beginning at the origin and counting up from it, you find the y–intercept to be at $(0,6)$. To find the slope, check the graph of m for rise over run, or in this case, negative rise over run (fall over run). The slope is $-\frac{3}{2}$.

Next you need to substitute the information you just collected into the standard equation for slope and y–intercept:

$$y = mx + b$$
$$y = \left(-\frac{3}{2}\right)x + (6)$$
$$y = -\frac{3}{2}x + 6$$

To put this equation into a more friendly looking form, you might multiply everything by 2. Then you would find:

$$y = -\frac{3}{2}x + 6$$

$$2\left(y = -\frac{3}{2}x + 6\right)$$

$$2y = -3x + 12$$

At what point will this graph have an x value of 12? You could read that as "What is the y–coordinate of the point on this graph where $x = 12$?" To find that, all you need to do is to substitute 12 for x and solve for y:

$$2y = -3x + 12$$

$$2y = -3(12) + 12$$

$$2y = -36 + 12 = -24$$

$$y = -12$$

The point is $(12, -12)$. At what point on this graph will it pass through $y = 18$? If you're thinking "I just have to substitute 18 for y in the equation and solve for x," then you have a good head on your shoulders.

So, the point's coordinates are $(-8, 18)$.

$$2y = -3x + 12$$

$$2(18) = -3x + 12$$

$$36 - 12 = -3x + 12 - 12$$

$$24 = -3x$$

$$x = -8$$

You've seen that given the slope and *y*–intercept of an equation, you can plot it or figure out the algebraic expression of that equation. In fact, if you know both the *y*–intercept and the *x*–intercept, you can draw the graph of the equation and figure its algebraic expression from there.

Look at the figure to the right. The difference in the coordinates of the two intercepts will tell you the slope.

You don't always need to know the *y*–intercept of an equation in order to figure out the equation or its graph. Knowing the coordinates of just one point on the equation's graph and its slope are all the information required when using the **Point-Slope method**.

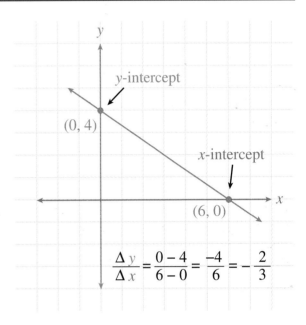

$$\frac{\Delta y}{\Delta x} = \frac{0-4}{6-0} = \frac{-4}{6} = -\frac{2}{3}$$

The Point-Slope formula is derived from the formula for slope at right:

$$m = \frac{y_2 - y_1}{x_2 - x_1}$$

By cross-multiplying, you get:

$$\frac{m}{1} = \frac{y_2 - y_1}{x_2 - x_1}$$
$$y_2 - y_1 = m\left(x_2 - x_1\right)$$

By altering that equation slightly, you derive the standard Point-Slope equation:

$$y - y_1 = m(x - x_1)$$

Consider a line with a slope of –2 and point (3, 5) on it. Consider the coordinates of the point to be the (x_1, y_1) in the preceding formula. Substituting these values into that formula, you get:

$$y - y_1 = m(x - x_1)$$
$$y - 5 = -2(x - 3)$$

Rearranging those terms to get the slope and intercept form, you find:

$$y - 5 = -2(x - 3)$$
$$y - 5 = -2x + 6$$
$$y - 5 + 5 = -2x + 6 + 5$$
$$y = -2x + 11$$

You can see its graph in this figure.

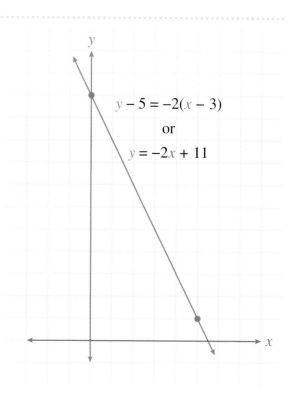

$$y - 5 = -2(x - 3)$$
or
$$y = -2x + 11$$

Now use the Point-Slope formula to write the equation of the line with slope 5 that contains the point $(-3, -6)$. Substituting those values into the formula, we get:

$$y - y_1 = m(x - x_1)$$
$$y - (-6) = 5(x - -3)$$

You have to be careful with those double negatives. Clear them and then collect like terms:

$$y - (-6) = 5(x - -3)$$
$$y + 6 = 5(x + 3)$$
$$y + 6 - 6 = 5x + 15 - 6$$
$$y = 5x + 9$$

The graph of the equation is shown in the figure to the right, as are both the Point-Slope and the slope and y–intercept equations.

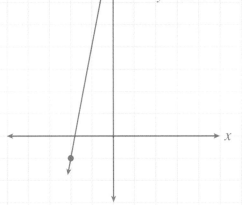

$$y - (-6) = 5[x - (-3)]$$

or

$$y = 5x + 9$$

TIP

If all you have are two points on a line, it's still quite simple to figure out that line's equation. Just find the difference between the two y–coordinates and the difference between the two x–coordinates in the same order.

Divide the first difference by the second, and you've found the slope. Now plug the newly found slope into the Point-Slope formula using the coordinates of either of the two points to write the equation of the line or to draw its graph.

Graphing Systems of Equations

You'll recall that Chapter 6 discussed solving systems of equations by addition/subtraction and by substitution, and I mentioned that a third method would be discussed later. Well, this is that later, and the third method for solving systems of equations is by *graphing*.

Consider the following system of equations:

$$3x + y = 13$$
$$3y - x = 9$$

To graph the first of the two equations, you can easily rewrite it in slope and y–intercept form, as $y = -3x + 13$. Its graph is shown in the figure to the right.

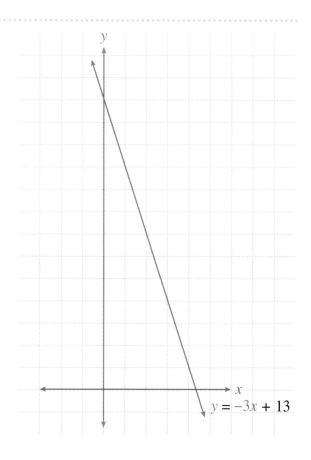

$y = -3x + 13$

The bottom equation rearranges as $3y = x + 9$, then:

$$3y = x + 9$$
$$\frac{3y}{3} = \frac{x + 9}{3}$$
$$y = \frac{1}{3}x + 3$$

That's graphed in this figure:

Where the two graphs cross is the solution to the system of equations. They cross at the point with coordinates $(3, 4)$. That means $x = 3$ and $y = 4$. Substitute those values into either or both of the equations and see for yourself.

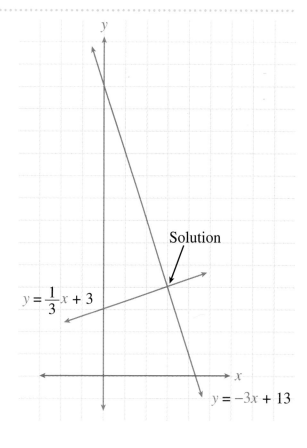

Solution

$y = \frac{1}{3}x + 3$

$y = -3x + 13$

Had those equations not lent themselves to being readily put into slope and *y*–intercept form, you could have plotted the graphs by substituting various values for *x* in each equation and then finding the value of *y* at that point, as you'll do later. Solve this pair of equations by graphing:

$$3x + 4y = -1$$
$$5y - 4x = 22$$

The trick is to select values for *x* that will result in whole number values for *y*. The equations are rewritten, and the following values worked:

Equation 1	Equation 2
$y = \dfrac{-1 - 3x}{4}$	$y = \dfrac{4x + 22}{5}$

x	y
5	−4
1	−1
−3	2

x	y
7	10
2	6
−3	2

The equations are graphed on the same set of axes in this figure. Where the two graphs cross is the solution to the system of equations. They cross at the point with coordinates (–3, 2). That means *x* = –3 and *y* = 2. Substitute those values into either or both of the equations and see for yourself. By the way, if you'd looked carefully at the tables you'd have seen the solution there as well. The point (–3, 2) is on both tables.

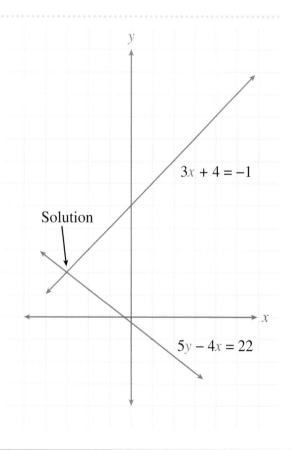

Practice Questions

1 In which quadrant would you find each of these points with the coordinates given?

a. $(-3, -5)$

b. $(7, -8)$

c. $(-5, 19)$

2 Who invented the coordinate system that bears his name?

3 Use substitution to graph the equation $6x + 3y = 27$.

4 Using substitution, graph the equation $4y - 12x + 16 = 0$.

5 Refer to this figure to state the slope and y-intercept of lines l and m.

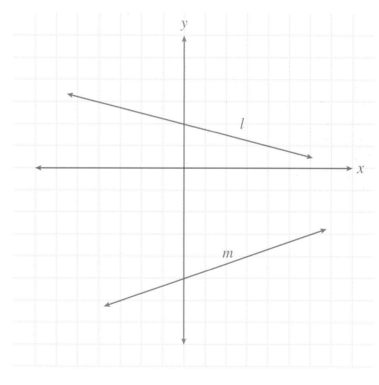

6 Refer to this figure to find the slope and *y*-intercept of lines *p* and *r*.

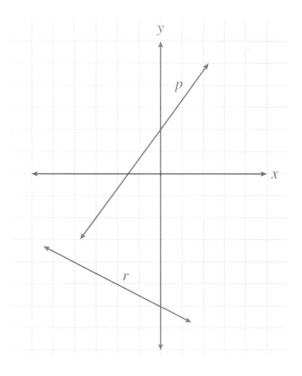

7 What is the slope of the line whose equation is $x = -8$?

8 What is the slope of the line whose equation is $y = -9$?

Use the slope and *y*–intercept method to graph the equations in questions 10–12.

9 $3x - 4y - 1 = 27$

10 $6x = 2y + 12$

11 What is the slope of a line that is parallel to $2y = -3x + 5$?

12 What is the slope of a line that is perpendicular to

a) $y = \frac{2}{3}x - 5$?

b) $3y = 7 - 12x$?

⑬ Referring to the figure at right, write the equation of line *a*.

⑭ Referring to the figure at right, write the equation of line *b*.

⑮ Referring to the figure at right, write the equation of line *c*.

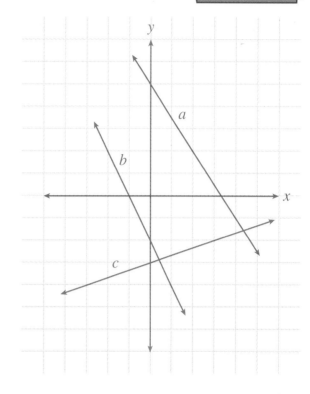

⑯ Write the equation of the line with a slope of 5 that passes through the point (–3, –4) in Point-Slope form.

⑰ Write the equation of the line with a slope of –4 that passes through the point (7, –9) in Point-Slope form.

⑱ Write the equation of the line with a slope of $\frac{4}{3}$ that passes through the point (–5, 8) in Point-Slope form.

Chapter Practice Answers

1 a. III, b. IV, c. II

2 Rene Descartes.

3 Here are four points, the top three of which I used in determining the graph.

x	y
2	5
1	7
0	9
−1	11

Other numbers are certainly possible, but your graph should look like the one in the figure below.

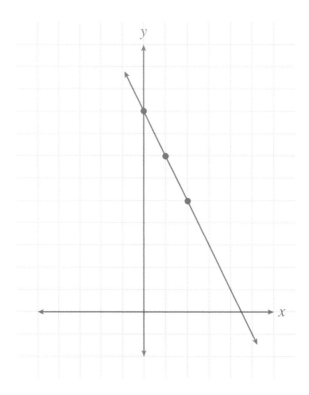

❹ Here are the three points that I used in determining the graph.

$$\begin{array}{c|c} x & y \\ \hline 1 & -1 \\ 0 & -4 \\ -1 & -7 \end{array}$$

Other numbers are certainly possible, but your graph should look like the one below.

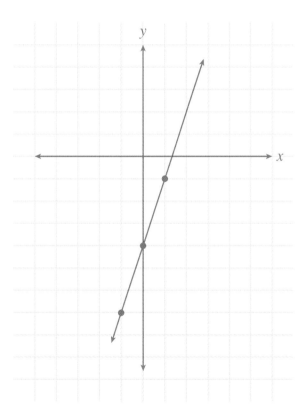

5 Line *l* has a slope of $-\dfrac{1}{4}$ and a *y*-intercept of 2. Line *m* has a slope of $\dfrac{1}{3}$ and a *y*-intercept of –5.

6 Line *p* has a slope of $\dfrac{4}{3}$ and a *y*-intercept of 2. Line *r* has a slope of $-\dfrac{1}{2}$ and a *y*-intercept of –6.

7 $x = -8$ is the equation of a vertical line. The slope of any vertical line is undefined.

8 $y = -9$ is the equation of a horizontal line. The slope of any horizontal line is 0.

9 First, the equation must be rewritten into the form $y = mx + b$:

$$3x - 4y - 1 = 27$$
$$3x - 3x - 4y - 1 + 1 = -3x + 27 + 1$$
$$-4y = -3x + 28$$
$$\frac{-4y}{-4} = \frac{-3x + 28}{-4}$$
$$y = \frac{3}{4}x - 7$$

Then the graph is drawn, as in the following figure.

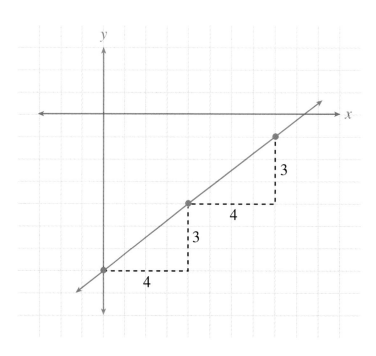

⑩ First, the equation must be rewritten into the form $y = mx + b$:

$$6x = 2y + 12$$
$$6x - 12 = 2y + 12 - 12$$
$$6x - 12 = 2y$$
$$\frac{2y}{2} = \frac{6x - 12}{2}$$
$$y = 3x - 6$$

Then the graph is drawn, as in this figure.

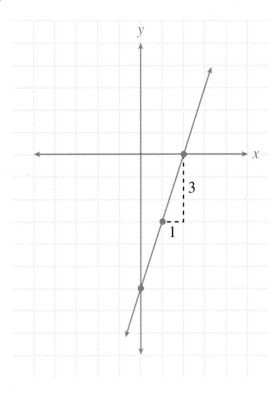

⑪ First find the slope:

$$2y = -3x + 5$$
$$\frac{2y}{2} = \frac{-3x + 5}{2}$$
$$y = -\frac{3}{2}x + \frac{5}{2}$$

Since the slope is $-\frac{3}{2}$, the slope of any parallel line must be the same.

⑫ The slope of a line that is perpendicular to another has as its slope the negative reciprocal of that line's slope.

a) $y = \frac{2}{3}x - 5$'s slope is $\frac{2}{3}$, so the slope of its perpendicular is $-\frac{3}{2}$

b) $3y = 7 - 12x$'s slope is -4, so the slope of its perpendicular is $\frac{1}{4}$.

The solutions to 15, 16, and 17 are all expressed in simplest form.

⑬ $m = -\frac{3}{2}$, $b = 5$, so substituting in the equation $y = mx + b$ you get:

$$y = -\frac{3}{2}x + 5$$

⑭ $m = -\frac{4}{3}$, $b = -2$, so substituting in the equation $y = mx + b$ you get:

$$y = -\frac{4}{3}x - 2$$

⑮ $m = \frac{1}{3}$, $b = -3$, so substituting in the equation $y = mx + b$ you get:

$$y = \frac{1}{3}x - 3$$

⑯ $y_1 - y = m(x_1 - x)$
$y - (-4) = m[x - (-3)]$
$y + 4 = 5(x + 3)$

⑰ $y_1 - y = m(x_1 - x)$
$y - (-9) = m(x - 7)$
$y + 9 = -4(x - 7)$

⑱ $y_1 - y = m(x_1 - x)$
$y - 8 = m(x - -5)$
$y - 8 = \frac{4}{3}(x + 5)$

chapter 9

Inequalities and Absolute Value

"All things being equal," is an expression you've heard and probably have used. The fact of the matter, however, is that all things are not equal; most are unequal, and therein lies the theme of this chapter. Mathematics allows for the unequalness of quantities through the mechanism of inequalities. In this chapter, you'll study all aspects of inequalities, including four different strategies for solving them as well as graphing them on both number lines and pairs of coordinate axes. You also look at the meaning of absolute value in considerably greater depth than when it was first introduced in Chapter 2. You'll examine solving both equations and inequalities that incorporate absolute value, trusting that you'll enjoy both equally.

Over the past few chapters you've become accustomed to working with equations—balances in which what is on one side of the equal sign is worth the same as what's on the other side. Ah, if only all things were created equal, but I'm sure you know that they are not. For those things, you have *inequalities*.

Inequalities come in four varieties:

    $<$, meaning less than
    $>$, meaning greater than
    $\leq$, meaning less than or equal to
    $\geq$, meaning greater than or equal to

The first two are self explanatory: $3 < 6$ means 3 is less than 6, and $7 > 5$ means 7 is greater than 5. They may apply to constants as well as to variables. The last two, however, might take a bit of exploring, since they apply primarily to variables. $5 \leq 7$ is an impossibility, since 5 is *always* less than 7 and can *never* be equal to it. Similarly, impossible is the inequality $8 \geq 6$, since 8 can *never* be equal to 6. That brings us to the two general types of inequality, *true and false*. The last two inequalities are examples of false ones.

$x \geq 4$ is a true inequality when $x = 4$ or anything greater than 4. Should $x = 3.9$ or anything less, that inequality becomes false. On the other hand, $x \leq 5$ is a true inequality when $x = 5$ or anything smaller than 5. Should $x = 5.1$ or anything greater, that inequality becomes false.

All of the preceding are examples of **simple inequalities**. Here's an example of a **compound** one: $5 \leq y \leq 9$. A **compound inequality** is read *from the middle, out*. That's going to make the color coding used unreliable. The foregoing says y is greater than or equal to 5, and it is less than or equal to 9. Any amount between 5 and 9 would qualify to make it a true statement. I know that's potentially confusing, but think about it. Since you're starting to read it from the middle, the first part is really the reverse of the symbol when read from the beginning.

What is the meaning of this statement? $11 \geq x \geq 8$

What whole numbers are in the answer set to the preceding inequality? The second question was thrown in to keep you from peeking at the answer to the first. It is read "x is less than or equal to 11 and greater than or equal to 8." Someone once suggested to view the $<$ and $>$ as an alligator's jaws with the smaller one looking to snap on the larger. Or, it might help to think of the arrowhead *always* pointing at the smaller number. The answer to the question about the answer set, by the way, is $\{8, 9, 10, 11\}$. If you think of the forward/backward logic of the way the symbols are read, you should be able to avoid (I hope) being confused, after all less than when read *left to right* becomes greater than when read from *right to left*.

Check out this statement: $6 > x \geq 12$

What is the meaning of it, and what is the answer set? It's probably easiest to start out with the answer set, which is $\{\ldots, -1, 0, 1, 2, 3, 4, 5, 12, 13, 14, \ldots\}$. The expression, when read from the center left says "x is less than 6 and greater than or equal to 12." That makes the answer set infinite in both the negative direction below 6 and in the positive direction 12 and above.

Inequalities are most often graphed on number lines. You're going to begin by graphing every inequality that was discussed in the last section.

For openers, here are the first two inequalities:

$$x \geq 4 \text{ and } x \leq 5$$

Notice that on both graphs there is a solid dot on the position corresponding to the number mentioned in the inequality. That solid dot represents the "equal to" part of the is greater than or equal to in a) and of the is less than or equal to in b).

a)

b)
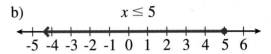

For contrast, look at the these graphs of $x > 4$ and $x < 5$. Notice the open dot on the 4 in figure a) and on the 5 in figure b). These indicate that those respective points are *not* included on the graphs.

a)
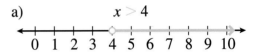

b)
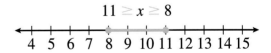

This graph is a bit different. It's of $11 \geq x \geq 8$. Notice that this graph is quite finite, with two solid dots as terminating points.

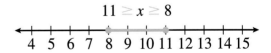

See how that contrasts with almost its opposite, $5 \leq y \leq 9$.

Good grief, the graph in this figure is identical to the graph in the preceding figure except for its position on the number line. Notice also that the variable's being y does not make the graph vertical. The Cartesian coordinate system has nothing to do with graphing these inequalities. Don't take that to mean that inequalities can't be graphed on axes, however; that's a different story that you're not yet ready to deal with.

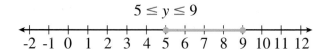

$$5 \leq y \leq 9$$

There are two more number-line graphs that you still need to look at. First look at the inequality $6 > x \geq 12$, as graphed here.

You looked at this inequality at the end of the last section, but look how graphing it almost brings it to life. You can see the separation between the part that is less than 6 and the part that is equal to or greater than 12. Also notice that the less than 6 part has an empty dot to show that 6 is not included on the graph.

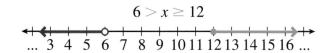

$$6 > x \geq 12$$

For the last number-line graph, consider $4 < w < 9$.

The thing that should be most noticeable about the graph depicted in this figure is that neither 4 nor 9 is included on the graph. For that reason, empty dots are at either end.

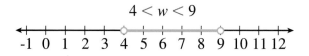

$$4 < w < 9$$

Solving inequalities is not very different from solving equations: In order to maintain the validity of the inequality, what is done on one side of the inequality sign must be likewise done on the other.

Look at a couple of instances, starting with this one:

$$y + 11 \leq 15$$

In order to get the variable y alone on one side of the inequality sign, you'll need to subtract 11 from the side that it's on, but if you do that, what else must you do?

$$y + 11 \leq 15$$
$$y + 11 - 11 \leq 15 - 11$$
$$y \leq 4$$

Hopefully, you said, "subtract 11 from the other side." And, so, you find that y is less than or equal to 4. Its graph is shown here.

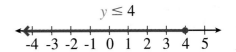

Try one more. Consider this:

$$x - 8 > -5$$

This time, you're going to need to add 8 to each side of the inequality.

$$x - 8 + 8 > -5 + 8$$
$$x > 3$$

You found that x is greater than 3. The graph of the solution is shown here.

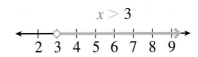

Note that this time the dot is open, indicating that 3 is not a part of the solution set.

Now I'm going to throw in one more little wrinkle. How do you suppose you'd solve this?

$$7 \geq x + 9$$

You could proceed here in two different ways, but I believe it will be easier to understand if you first get the variable by itself, so subtract 9 from each side.

$$7 \geq x + 9$$
$$7 - 9 \geq x + 9 - 9$$
$$-2 \geq x$$

Now you know that that answer doesn't look quite right, and the reason that it doesn't is because the variable is on the wrong side of the inequality sign—that is, on the right, rather than the left. Look at the expression $-2 \geq x$ and say what it means: Negative 2 is greater than or equal to x. Isn't that really saying the same thing as x is equal to or less than negative 2?

Of course it does, and that's the answer. Looks better now, doesn't it?

$$x \leq -2$$

I mentioned earlier that there were two ways to solve $7 \geq x + 9$. Now, knowing what you know, look at the other way. Turn it around from the start and then subtract 9 from each side:

$$7 \geq x + 9$$
$$x + 9 \leq 7$$
$$x + 9 - 9 \leq 7 - 9$$
$$x \leq -2$$

The graph of the answer is shown here.

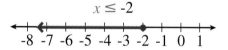

$x \leq -2$

-8 -7 -6 -5 -4 -3 -2 -1 0 1

You probably think that you already know how to solve inequalities by multiplying and dividing—and to some extent you'd be correct. But, as you may have figured out from the last section, there are a couple of exceptions.

First, here is one example that you probably could have anticipated:

$$5x \leq 30$$

If you're thinking the solution is to divide both sides of the inequality by 5, you'd be absolutely correct.

$$5x \leq 30$$
$$\frac{5x}{5} \leq \frac{30}{5}$$
$$x \leq 6$$

And, so, you find that x is *less than or equal to* 6. Now, here's an exception to the expected. Look at this:

$$-4y > 12$$

You're probably saying to yourself, "self, what's the problem? All I need to do is to divide through by -4," and up to a point you'd be correct.

$$-4y > 12$$
$$\frac{-4y}{-4} > \frac{12}{-4}$$

Guess what? You've now reached that point. Because dividing by -4 is going to change the sign of the variable, something else must also be changed.

$$\frac{-4y}{-4} > \frac{12}{-4}$$
$$y < 3$$

That's right, is greater than changed to is less than. Would you like a moment to think about that? If $-4y$ is greater than 12, then $-y$ is greater than 3, so wouldn't $-y$'s opposite have an opposite relationship with 3? Hopefully, that makes it a bit clearer.

TIP

The important thing to remember is if the variable's sign changes, the inequality sign changes!

Try one more:

$$-6x \leq 18$$

What do you think? Do you have a solution yet? Well, I hope it looks something like this:

$$-6x \leq 18$$
$$\frac{-6x}{-6} \leq \frac{18}{-6}$$
$$x \geq 3$$

Did you get that?

Look at a couple of solutions by multiplication. First:

$$\frac{4}{3}z < -12$$

Do you know how to solve this inequality? If you're thinking "multiply each side by $\frac{3}{4}$," then your thinking cap is evidently in good working order.

$$\frac{4}{3}z < -12$$
$$\frac{3}{4} \cdot \frac{4}{3}z < -12 \cdot \frac{3}{4}$$
$$z < -9$$

That's because $-\frac{36}{4} = -9$, just in case you got confused for a moment.

Now look at one last example:

$$-\frac{3}{2}x \leq 9$$

If you want to try figuring out the solution for yourself, don't look any further down the page.

Ready? All right, then, here's the solution. I'm sure you figured out the fact that both sides need to be multiplied by $-\frac{2}{3}$, but did you consider the consequence?

$$-\frac{3}{2}x \leq 9$$
$$\left(-\frac{2}{3}\right)\left(-\frac{3}{2}x\right) \leq 9\left(-\frac{2}{3}\right)$$
$$x \geq -\frac{18}{3}$$
$$x \geq -6$$

You did remember to reverse the inequality sign, I'm sure.

The discussion touched upon absolute value in Chapter 2. The distance from 0 that a number is on the number line is known as its absolute value. It is a concept that is one of the underpinnings of working with equations and inequalities of a higher order than you've dealt with to this point.

If you look at this figure, you'll see two points marked with A. They are marked that way because they both have the same absolute value, 7. The points marked with a B both have an absolute value of 4; those marked C have an absolute value of 2.

$$\begin{array}{ccccccccc} A & & B & C & & C & B & & A \end{array}$$

-8 -7 -6 -5 -4 -3 -2 -1 0 1 2 3 4 5 6 7 8

Read the equation to the right as "the absolute value of negative 3 equals 3." Of course, the absolute value of positive 3 also equals 3.

$$|-3| = 3$$

Now consider this:

$$|w| = 7$$

It reads, "the absolute value of w equals 7." What is the value of w? Does that sound like a foolish question? Were you tempted to pick the obvious answer, 7? If you did, you're half right. The value of w is $\{-7, 7\}$. That's right, w has two different values, since both 7 and -7 fit the criterion for the absolute value of w. $|7| = 7$ and $|-7| = 7$. This should help to caution you to proceed carefully when it comes to solving equations and inequalities involving absolute value.

Solving Equations Containing Absolute Value

Look at this equation containing an absolute value:

$$3|x - 5| + 9 = 18$$

The strategy to follow is to go about getting the absolute value by itself on one side of the equal sign by first adding or subtracting (as needed) and then multiplying or dividing (if appropriate). In this case you'll start by subtracting 9 from each side.

$$3|x - 5| + 9 - 9 = 18 - 9$$
$$3|x - 5| = 9$$

Next, divide each side by the 3 that the absolute value is currently multiplied by.

$$\frac{3|x-5|}{3} = \frac{9}{3}$$
$$|x-5| = 3$$

The final step is to remove the absolute value sign and set the quantity in the bracket equal to positive 3 and negative 3.

$$
\begin{array}{r}
x - 5 = 3 \\
+5 \quad +5 \\
\hline
x = 8
\end{array}
\qquad
\begin{array}{r}
x - 5 = -3 \\
+5 \quad +5 \\
\hline
x = 2
\end{array}
$$

So $x = \{2, 8\}$ is the solution set.

Try one more absolute value equation:

$$\frac{2}{3}|r+6| - 7 = 9$$

Following the already established strategy, first add 7 to each side.

$$\frac{2}{3}|r+6| - 7 + 7 = 9 + 7$$
$$\frac{2}{3}|r+6| = 16$$

To finish collecting terms (a way of saying get all of one type of term on one side and all constants on the other), *multiply both sides* by the reciprocal of $\frac{2}{3}, \frac{3}{2}$.

$$\frac{3}{2} \cdot \frac{2}{3}|r+6| = \frac{3}{2} \cdot 16$$
$$|r+6| = 24$$

Finally, remove the absolute value sign and set the quantity in the bracket equal to positive 24 and negative 24.

$$
\begin{array}{r}
r + 6 = 24 \\
-6 \quad -6 \\
\hline
r = 18
\end{array}
\qquad
\begin{array}{r}
r + 6 = -24 \\
-6 \quad -6 \\
\hline
r = -30
\end{array}
$$

The solution set is $\{18, -30\}$.

Solving Inequalities Containing Absolute Value

As you may have surmised, very little difference exists between the strategies used to solve equations containing absolute value and those used to solve inequalities containing the same. The major difference will be in the solutions and the different ways of expressing and graphing them. Since there are going to be two different solutions, when you set the inequality <> 3, the negative of the solution, that sign must be reversed. That is, when the inequality $|x| > 3$, $x > 3$ is negated and it becomes $x < -3$. That is, x is both greater than 3 and less than -3.

Try a couple of examples, so you can get used to the procedure. Solve and graph the following:

$$2|x-2| > 8$$

. . . .

First, you need to isolate the absolute value by dividing both sides by 2.

$$2|x-2| > 8$$
$$\frac{2|x-2|}{2} > \frac{8}{2}$$
$$|x-2| > 4$$

. . . .

Now separate the absolute value to cover both positive 4 and negative 4.

$$x-2 > 4 \qquad\qquad x-2 < -4$$
$$x-2+2 > 4+2 \qquad x-2+2 < -4+2$$
$$x > 6 \qquad\qquad x < -2$$

. . . .

The solution is graphed in the figure to the right; note that neither the point 6 nor the point −2 is included.

$$x > 6$$

$$\xleftarrow{\quad} \; \overset{\circ}{|} \; \underset{\text{-3 -2 -1 0 1 2 3 4 5 6 7 ...}}{| \; | \; | \; | \; | \; | \; | \; | \; | \;} \; \overset{\circ}{|} \; \xrightarrow{\quad}$$

$$x < 2$$

. . . .

Now work one in which the points will be included. Solve and graph this one:

$$|n+3| + 5 \le -2$$

Understanding
Absolute Value *(continued)*

To get the absolute value term by itself, you'll need to subtract 5 from both sides of the inequality.

$$|n + 3| + 5 \leq -2$$
$$|n + 3| + 5 - 5 \leq -2 - 5$$
$$|n + 3| \leq -7$$

Here comes the tricky part. -7 is already negative, so are you going to have to reverse the inequality sign? Take a moment or two to think about that. The answer is of course you do, but you'll be reversing it for the positive 7 you're about to create.

$$n + 3 \geq 7 \qquad\qquad n + 3 \leq -7$$
$$n + 3 - 3 \geq 7 - 3 \qquad\qquad n + 3 - 3 \leq -7 - 3$$
$$n \geq 4 \qquad\qquad n \leq -10$$

The solution is graphed in the figure to the right; note that both the point 4 and the point -10 are included.

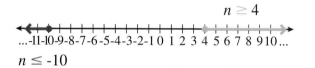

$$n \geq 4$$

...-11-10-9-8-7-6-5-4-3-2-1 0 1 2 3 4 5 6 7 8 9 10...

$$n \leq -10$$

When a line is plotted on a pair of coordinate axes, it divides the graph into two half-planes. If the line is vertical, then the half-planes will be to the right and the left of that boundary; a horizontal or oblique line will separate the graph into the half-plane above it and the half-plane below it. When an inequality is graphed on coordinate axes, the result is always a half-plane.

If it seems to you that what is on one side of the line must be greater than what is on the other except when the line passes through the origin, then you are losing sight of the fact that the plane itself is infinite in nature. Since what is on either side of the plotted line is also infinite, then those half-planes are by extension equal in size.

When an inequality is a $<$ or a $>$, then the graph will be what is known as an **open half-plane**. That means that the boundary line is not included in the solution set. To indicate that, the line is drawn dotted. To see how this works, graph the inequality:

$$y > x + 4$$

First you find a few values by substitution:

x	y
0	4
−2	2
−4	0

I chose those values because they were convenient for me, but you could have substituted any value for x to solve for y. You also could have drawn it using the slope and y-intercept method. The result is shown in the figure to the right. Note that the open half-plane is above the boundary line, which is dotted to show its exclusion.

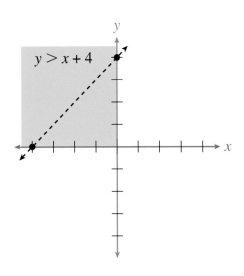

When an inequality is a $\leq$ or a $\geq$, then the graph will be what is known as a **closed half-plane**. That means that the boundary line **is** included in the solution set. To indicate that, the line is drawn solid. To see how that works, graph this inequality:

$$y \leq x - 3$$

If you consider this equation to be in the slope and y-intercept form, then its slope is the coefficient of x, namely 1, and the y-intercept is $(-3, 0)$. Otherwise, you could plot a few points by the substitution method.

x	y
2	-1
0	-3
-2	-5

The solution is shown in the figure to the right. Note that the closed half-plane is below the boundary line, which is solid to show its inclusion.

You'll get the opportunity to work on some of these in the Chapter Practice, which is coming up next.

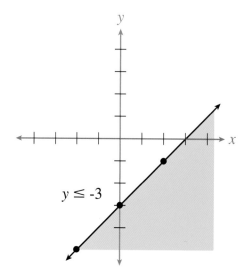

Unless otherwise specified, restrict your responses to the realm of integers.

1 a. What is the answer set to the following?

  b. How is it read?

$$6 \leq x \leq 9$$

2 a. What is the answer set to the following?

  b. How is it read?

$$4 > x \geq 12$$

3 Solve the following and graph it on a number line:

$$6 \geq x + 8$$

4 Solve the following and graph it on a number line:

$$4n + 3 \geq 31$$

5 Solve the following and graph it on a number line:

$$-6y > 24$$

6 Solve the following and graph it on a number line:

$$-\frac{3}{4}x \geq 9$$

7 Find the value(s) of r that satisfy the equation:

$$2|r - 4| + 9 = 11$$

8 Find the value(s) of n that satisfy the equation:

$$\frac{4}{3}|n+3| - 7 = 21$$

9 Solve and graph the following on a number line:

$$3|x-5| > 12$$

10 Solve and graph the following on a number line:

$$4|y-5| - 12 < 20$$

11 Solve and graph the following on a number line:

$$|n+3| - 5 \leq -2$$

12 Solve and graph the following on a number line:

$$|x-3| + 5 \geq 4$$

Graph the following inequalities on a pair of coordinate axes.

13 $x \geq 5$

14 $3y - 6x < 9$

15 $y - 5x \geq 4$

Chapter Practice Answers

1 a. {6, 7, 8, 9}

b. The expression, when read from the center left says *"x is greater than or equal to 6 and less than or equal to 9."*

2 a. {. . . , –1, 0, 1, 2, 3, 12, 13, 14, . . .}

b. The expression, when read from the center left, says *"x is less than 4 and equal to 12 or greater."* That makes the answer set infinite in both the negative direction below 4 and in the positive direction 12 and above.

3 $x \le -2$

The procedure followed here is not the only possible way to solve this, but the result will be the same. First you get the variable by itself, by subtracting 8 from each side:

$$6 \ge x + 8$$
$$6 - 8 \ge x + 8 - 8$$
$$-2 \ge x$$

Now that answer doesn't look quite right, and the reason is that the variable is on the wrong side of the inequality sign—that is the right, rather than the left. Look at the expression $-2 \ge x$ and say what it means: Negative 2 is greater than or equal to *x*. Isn't that really saying the same thing as *x* is equal to or less than negative 2?

$$x \le -2$$

The graph is shown here.

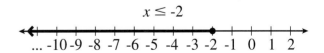

$x \le -2$

$$\cdots \; -10 \; -9 \; -8 \; -7 \; -6 \; -5 \; -4 \; -3 \; -2 \; -1 \; 0 \; 1 \; 2$$

4 $n \geq 7$

First, subtract three from both sides; then divide both sides by 4:

$$4n + 3 \geq 31$$
$$4n + 3 - 3 \geq 31 - 3$$
$$4n \geq 28$$
$$\frac{4n}{4} \geq \frac{28}{4}$$
$$n \geq 7$$

So, find that n is greater than or equal to 7. The graph is shown here.

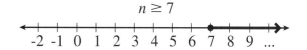

5 $y < 4$

The solution begins in a straightforward way. Both sides need to be divided by 6, but . . .

$$-6y > 24$$
$$\frac{-6y}{-6} > \frac{24}{-6}$$
$$y < 4$$

Because the sign of the variable has reversed, the inequality sign must also be reversed. Notice that on the graph you should have made an empty dot at 4, since y is less than 4.

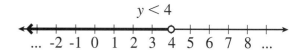

❻ $x \leq 12$

To solve the equation, both sides must be multiplied by the reciprocal of $\frac{3}{4}$, $\frac{4}{3}$. Furthermore, since the variable was formerly preceded by a – sign, the inequality sign must be reversed.

$$-\frac{3}{4}x \geq 9$$
$$-\frac{3}{4}\left(\frac{4}{3}\right)x \geq 9\left(\frac{4}{3}\right)$$
$$-x \geq \frac{36}{3}$$
$$x \leq 12$$

The graph of the solution is shown here.

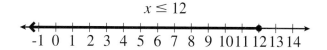

$x \leq 12$

-1 0 1 2 3 4 5 6 7 8 9 10 11 12 13 14

❼ $r = \{3, 5\}$

You need to get the absolute value by itself on one side of the equals sign, so first subtract 9 from both sides and then divide by 2:

$$2|r-4|+9 = 11$$
$$2|r-4|+9-9 = 11-9$$
$$2|r-4| = 2$$
$$\frac{2|r-4|}{2} = \frac{2}{2}$$
$$|r-4| = 1$$

The final step is to remove the absolute value sign and set the quantity in the bracket equal to positive 1 and negative 1.

$$
\begin{array}{ll}
r - 4 = 1 & r - 4 = -1 \\
 +4 +4 & +4 +4 \\
\hline
r = 5 & r = 3
\end{array}
$$

So $r = \{3, 5\}$ is the solution set.

8 $\{18, -24\}$

To solve this equation, first add to both sides; then multiply by the reciprocal of the fraction $\frac{4}{3}$.

$$\frac{4}{3}|n + 3| - 7 = 21$$

$$\frac{4}{3}|n + 3| - 7 + 7 = 21 + 7$$

$$\frac{4}{3}|n + 3| = 28$$

$$\frac{3}{4} \cdot \frac{4}{3}|n + 3| = \frac{3}{4} \cdot 28$$

$$|n + 3| = \frac{3}{1} \cdot \frac{\cancel{28}^{7}}{\cancel{4}}$$

$$|n + 3| = 21$$

Finally, remove the absolute value brackets and set the left side equal to positive and negative 21.

$$
\begin{array}{ll}
n + 3 = 21 & n + 3 = -21 \\
 -3 -3 & -3 -3 \\
\hline
n = 18 & n = -24
\end{array}
$$

9 $x > 9$, $x < 1$ or $x < 1 > 9$

To get the absolute value term by itself, you'll need to divide both sides of the inequality by 3:

$$3|x - 5| > 12$$
$$\frac{3|x - 5|}{3} > \frac{12}{3}$$
$$|x - 5| > 4$$

Now comes the tricky part. Remove the bracket and rewrite for both positive and negative with the sign reversed:

$$
\begin{array}{ccc}
|x - 5| > 4 & \text{or} & |x - 5| < -4 \\
x - 5 + 5 > 4 + 5 & & x - 5 + 5 < -4 + 5 \\
x > 9 & & x < 1
\end{array}
$$

The solution is graphed in the following figure; note that both the point 9 and the point 1 are *not* included.

$n < 1$ $n > 9$

... -3 -2 -1 0 1 2 3 4 5 6 7 8 9 10 11 ...

10 $-3 < y < 13$

To get the absolute value term by itself, you'll need to add 12 to both sides of the inequality and then divide both sides by 4:

$$4|y - 5| - 12 < 20$$
$$4|y - 5| - 12 + 12 < 20 + 12$$
$$4|y - 5| < 32$$
$$\frac{4|y - 5|}{4} < \frac{32}{4}$$
$$|y - 5| < 8$$

Now comes the cool part. Remove the bracket and rewrite for both positive and negative with the sign reversed:

$$|y-5| < 8 \qquad \text{or} \qquad |y-5| > -8$$
$$y-5+5 < 8+5 \qquad\qquad y-5+5 > -8+5$$
$$y < 13 \qquad\qquad\qquad y > -3$$
$$-3 < y < 13$$

The solution is graphed in the following figure; note that both the point −3 and the point 13 are *ex*cluded.

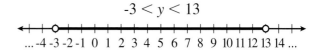

$-3 < y < 13$

... -4 -3 -2 -1 0 1 2 3 4 5 6 7 8 9 10 11 12 13 14 ...

⑪ $-6 \leq n \leq 0$

To get the absolute value term by itself, you'll need to add 5 to both sides of the inequality:

$$|n+3| - 5 \leq -2$$
$$|n+3| - 5 + 5 \leq -2 + 5$$
$$|n+3| \leq 3$$

Next remove the bracket and rewrite for both positive and negative with the sign reversed:

$$|n+3| \leq 3 \qquad\qquad |n+3| \geq -3$$
$$n+3-3 \leq 3-3 \qquad n+3-3 \geq -3-3$$
$$n \leq 0 \qquad\qquad\qquad n \geq -6$$

The solution is graphed in the following figure; note that both the point 0 and the point −6 are included.

$-6 \leq m \leq 0$

... -7 -6 -5 -4 -3 -2 -1 0 1 2 3

⑫ $2 \leq x \leq 4$

To get the absolute value term by itself, you'll need to subtract 5 from both sides of the inequality:

$$|x-3|+5 \geq 4$$
$$|x-3|+5-5 \geq 4-5$$
$$|x-3| \geq -1$$

Next remove the bracket and rewrite for both positive and negative with the sign reversed:

$$|x-3| \geq -1 \qquad\qquad |x-3| \leq 1$$
$$x-3+3 \geq -1+3 \qquad x-3+3 \leq 1+3$$
$$x \geq 2 \qquad\qquad\qquad x \leq 4$$

The solution is graphed in the following figure; note that both the point 4 and the point 2 are included.

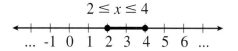

⑬ First graph the line $x = 5$. The solution is a closed vertical half plane, so the line is solid. Finally, decide which side to shade.

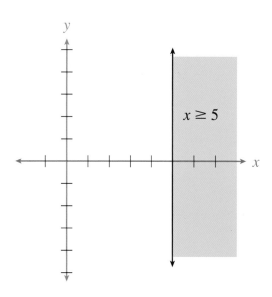

⑭ First graph the line $3y - 6x = 9$, which becomes $y = 2x + 3$. The solution is an open half plane, so the line is dotted. Finally, decide which side to shade.

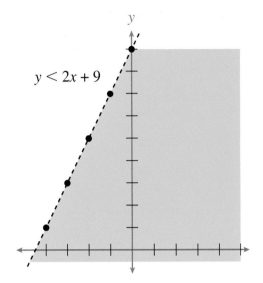

$y < 2x + 9$

⑮ First graph the line $y - 5x = 4$, which becomes $y = 5x + 4$. The solution is a closed half plane, so the line is solid. Finally, decide which side to shade.

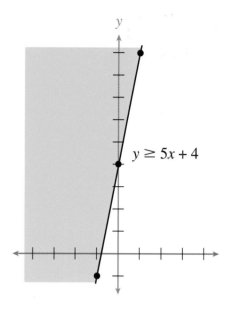

$y \geq 5x + 4$

chapter 10

Algebraic Fractions

You already dealt with fractions in Chapter 3. Those fractions involved signed numbers. So what makes algebraic fractions any different? Algebraic fractions contain an algebraic expression in the numerator, the denominator, or both. You may be surprised to learn that certain restrictions apply to algebraic fractions, and those restrictions stem from the fact that division by zero is not allowed. Look at this fraction:

$$\frac{5}{x}$$

What restriction is there on the value of x? The restriction is that x cannot be = to zero; that statement is abbreviated as $x \neq 0$. Now consider this new fraction:

$$\frac{y}{w-4}$$

What restrictions are there on the values of w and y? There are no restrictions on the value of y. However, $w \neq 4$! (That is w cannot equal 4.) That would make the denominator = 0, which is the same as dividing by 0. This is only a small fraction of what you're going to study in this chapter.

You've looked at some of the restrictions on algebraic fractions during the chapter introduction. Here, you're going to pursue that a bit further.

Look at the following fractions:

a) $\dfrac{2}{z+3}$

Just in case it was not clear to you in the chapter introduction, any fraction may be considered to be a division, with the denominator being the divisor and the numerator being the dividend, as was noted in Chapter 3. Since division by 0 is illegal—actually undoable—restrictions are placed on what the values of the variables in a denominator can be. What can z not be in a)? To answer that question, you need to ask yourself what when added to 3 would make 0? The answer is -3; z may not be equal to -3 ($z \neq -3$).

Now look at b). What restrictions are there on the values of x and y? Since one is being subtracted from the other, they may not both have the same value ($x \neq y$). May one of them be 0? Sure, as long as both are not.

b) $\dfrac{5}{x-y}$

Now take a look at c). It is very different from any of the fractions you have looked at before. In the first place, it contains an exponential term; secondly, the two quantities in the denominator are multiplied together. If either one of those terms were equal to 0, the entire denominator's value would become 0. Therefore, ($a \neq 0$) **and** ($x \neq 0$). Keep these restrictions in mind as you work with algebraic fractions.

c) $\dfrac{4}{a^2 x}$

Simplifying Algebraic Fractions

Some books use the expression *reducing* fractions. That is a misnomer, in that it suggests that the fraction is being *made smaller*. When simplifying a fraction, the terms in which it is expressed may be *reduced*, but the size of the fraction—its value—doesn't change. For that reason, you'll never see *reduce* used again with respect to fractions in this book.

To simplify an algebraic fraction, look for *common factors* in the numerator and denominator that may be divided out, or cancelled. Here's an example:

$$\frac{5x^2y}{10xy^2} = \frac{\cancel{5}^1 x^{\cancel{2}1} \cancel{y}^1}{\cancel{10}_2 \cancel{x}_1 y^1} = \frac{x}{2y} \text{ or } \frac{1}{2} \cdot \frac{x}{y}$$

Note that 5s, xs, and ys in both numerator and denominator could be factored (divided) out. Sometimes it is not so evident what is the same until some preliminary factoring is done, as in the following:

$$\frac{4a+8}{5a+10} = \frac{4(a+2)}{5(a+2)} = \frac{4\cancel{(a+2)}}{5\cancel{(a+2)}} = \frac{4}{5}$$

Factoring both numerator and denominator yields the common factor, $a + 2$. Sometimes, it's not as obvious as that.

Here's a for instance:

$$\frac{x^2+2x+1}{x^2-1}$$

If you thought that you factored polynomials in Chapter 7 just for fun, think again:

$$\frac{x^2+2x+1}{x^2-1} = \frac{(x+1)(x+1)}{(x+1)(x-1)} = \frac{\cancel{(x+1)}(x+1)}{\cancel{(x+1)}(x-1)} = \frac{x+1}{x-1}$$

And that completes the story.

None of the arithmetic operations with algebraic fractions is any different from the same operation with common fractions except that algebraic expressions are involved. That involvement, however, can make things rather complicated. When multiplying, it is best to write the results in parenthetical form and save any combining to be done for the very end. That greatly helps to ensure that the fraction is in simplest form. It's good policy to try to simplify the fractions by factoring and/or canceling before multiplying the resultant numerators and denominators.

Here are two simple examples:

$$\frac{2a}{5} \cdot \frac{3b}{7} = \frac{2 \cdot 3 \cdot a \cdot b}{5 \cdot 7} = \frac{6ab}{35}$$

$$\frac{3x}{4y^2} \cdot \frac{2y}{5x} = \frac{3\overset{1}{\cancel{x}}}{\underset{2}{\cancel{4}}\cancel{y}^{21}} \cdot \frac{\overset{1}{\cancel{2}}\overset{1}{\cancel{y}}}{5\cancel{x}_1} = \frac{3 \cdot 1 \cdot 1 \cdot 1}{2 \cdot 5 \cdot 1 \cdot y} = \frac{3}{10y}$$

These two examples involved monomials, but the main strategies used will apply to binomials and polynomials as well. Check out this one. It may not be evident why the colors are used as they are, but it will become apparent as factoring occurs.

$$\frac{x+1}{2x-2} \cdot \frac{3x-3}{4x+4} = \frac{(x+1)}{2(x-1)} \cdot \frac{3(x-1)}{4(x+1)}$$

Now there's some canceling that can be done, to get the result.

$$\frac{\overset{1}{\cancel{x+1}}}{2\cancel{(x-1)}_1} \cdot \frac{3\overset{1}{\cancel{(x-1)}}}{4\cancel{(x+1)}_1} = \frac{1 \cdot 3 \cdot 1}{2 \cdot 1 \cdot 4 \cdot 1} = \frac{3}{8}$$

Here's a final example of multiplying algebraic fractions. Check out the following:

$$\frac{24y^3(y^2-4y+4)}{25(x+3)} \cdot \frac{10x(x^2+6x+9)}{21y(y-2)(y^2-4y+4)}$$

The colors should make it easier to see which factors are going to cancel with which.

$$\frac{24y^3(y^2-4y+4)}{25(x+3)} \cdot \frac{10x(x+3)(x+3)}{21y(y-2)(y^2-4y+4)}$$

Factoring $x^2 + x + 9$ helps you to find other things to cancel.

$$\frac{\overset{8}{24}y^{3\,2}}{\underset{5}{25}(x+3)} \cdot \frac{\overset{2}{10}x(x+3)(x+3)}{\underset{7}{21}y(y-2)}$$

Finally, multiply.

And leave it in that factored form.

$$\frac{8y^2}{5} \cdot \frac{2x(x+3)}{7(y-2)} = \frac{16xy^2(x+3)}{35(y-2)}$$

Dividing Algebraic Fractions

Dividing algebraic fractions is accomplished by *changing the divisor to its reciprocal* **and then multiplying. You might recall it as** *invert the second fraction* **and multiply.**

Consider this example. Color is used to distinguish numerator from denominator.

$$\frac{3}{2x} \div \frac{y}{4} = \frac{3}{2x} \cdot \frac{4}{y}$$

Next, cancel if possible and multiply. Notice that the final product contains elements from the original numerator and denominator.

$$\frac{3}{\cancel{2}x} \cdot \frac{\cancel{4}^2}{y} = \frac{6}{xy}$$

Notice that the time for canceling—if the opportunity exists at all—is *after* changing the divisor to its reciprocal and *before* multiplying.

Let's try a slightly more complex algebraic fraction division. Use the same color scheme to distinguish numerator from denominator.

$$\frac{8x+4}{3} \div \frac{2x+1}{6} = \frac{8x+4}{3} \cdot \frac{6}{2x+1}$$

Change the divisor to its reciprocal and turn the division into a multiplication. Now it's time to see whether you can do some simplifying.

$$\frac{8x+4}{3} \cdot \frac{6}{2x+1} = \frac{4(2x+1)}{\cancel{3}_1} \cdot \frac{\cancel{6}^2}{2x+1}$$

And finally, multiply to get a grand total of 8.

$$\frac{4\cancel{(2x+1)}^1}{1} \cdot \frac{2}{\cancel{2x+1}} = \frac{4 \cdot 1 \cdot 2}{1} = 8$$

Try one more algebraic fraction division—a complex one. Stick with the same color scheme.

$$\frac{18\left(a^{2}+6a+9\right)}{5a^{2}bc^{3}}\div\frac{12\left(a+3\right)\left(b^{2}-4b+4\right)}{25abc}$$

First, change the divisor to its reciprocal and turn the division into a multiplication. Then see whether you can do some simplifying.

$$\frac{18\left(a^{2}+6a+9\right)}{5a^{2}bc^{3}}\cdot\frac{25abc}{12\left(a+3\right)\left(b^{2}-4b+4\right)}=\frac{\overset{3}{\cancel{18}}\left(\cancel{a+3}\right)\left(a+3\right)}{\underset{1}{\cancel{5}}a^{2}bc^{3}}\cdot\frac{\overset{5}{\cancel{25}}\,abc}{\underset{2}{\cancel{12}}\left(\cancel{a+3}\right)\left(b-2\right)\left(b-2\right)}$$

$$\frac{3\left(a+3\right)}{a^{1}\cancel{b}c^{2}}\cdot\frac{5\cancel{abc}}{2\left(b-2\right)\left(b-2\right)}=\frac{3\left(a+3\right)}{ac^{2}}\cdot\frac{5}{2\left(b-2\right)\left(b-2\right)}$$

Finally, multiply to get:

And that's what I call a "long division."

$$\frac{3\left(a+3\right)}{ac^{2}}\cdot\frac{5}{2\left(b-2\right)\left(b-2\right)}=\frac{15\left(a+3\right)}{2ac^{2}\left(b-2\right)^{2}}$$

If you've been wondering why I saved addition and subtraction of algebraic fractions for after you learned about multiplying and dividing them, the answer is simple: *It's harder!* If you recall, in order for fractions to be addable or subtractible, those fractions *must have the same denominators*.

Check out the following two fractions:

The denominator on the left has a 3, an x, and a y^2. The one on the right has a 2, an x^2, and a y. Each contains a constant, an x, and a y. But the constants are different, and so are the orders (or the degrees) of the variables. What you need here is a common denominator. The **least common denominator** (**LCD**) is $6x^2y^2$. Go ahead and check it out for yourself; don't take my word for it.

$$\frac{7}{3xy^2} + \frac{5}{2x^2y}$$

Now, to change the two original denominators to the LCD, you have to divide the LCD by each original denominator and multiply its numerator by the result. For example:

$$6x^2y^2 \div 3xy^2 = 2x \quad \text{and} \quad 6x^2y^2 \div 2x^2y = 3y$$

$$2x \cdot 7 = 14x \qquad\qquad 3y \cdot 5 = 15y$$

$$\frac{7}{3xy^2} = \frac{14x}{6x^2y^2} \qquad\qquad \frac{5}{2x^2y} = \frac{15y}{6x^2y^2}$$

Finally, add them together.

$$\frac{14x}{6x^2y^2} + \frac{15y}{6x^2y^2} = \frac{14x + 15y}{6x^2y^2}$$

Now wasn't that a barrel of laughs! If there was anything you didn't follow, go back and check each part again. They should fall into pace.

Ready for a subtraction? (That was a rhetorical question.) Check this one out:

$$\frac{5x}{4y} - \frac{2y}{5x}$$

What is the least common denominator? If you said $20xy$, you are correct. Divide that by each existing denominator and multiply the result by the old numerators to find what the new numerators will be.

$$20xy \div 4y = 5x \quad \text{and} \quad 20xy \div 5x = 4y$$

$$5x \cdot 5x = 25x^2 \qquad\qquad 4y \cdot 2y = 8y^2$$

$$\frac{5x}{4y} = \frac{25x^2}{20xy} \qquad\qquad \frac{2y}{5x} = \frac{8y^2}{20xy}$$

Adding and Subtracting Algebraic Fractions *(continued)*

Now rewrite the subtraction and solve it.

$$\frac{25x^2}{20xy} - \frac{8y^2}{20xy} = \frac{25x^2 - 8y^2}{20xy}$$

Sometimes to find the LCD it is necessary to factor both denominators, as is the case here:

$$\frac{3x}{x^2 - 16} + \frac{4}{x^2 + 8x + 16} = \frac{3x}{(x-4)(x+4)} + \frac{4}{(x+4)^2}$$

So what is the least common denominator? It has to be divisible by $(x-4)$ once and by $(x+4)$ twice. That makes it:

$$(x-4)(x+4)(x+4), \text{ or } (x-4)(x+4)^2$$

Divide that by each existing denominator and multiply the result by the old numerators to find what the new numerators will be.

Now you're ready to add.

$$\frac{(x-4)(x+4)^2}{(x-4)(x+4)} = (x+4)$$

$$\frac{3x^2 + 12x}{(x-4)(x+4)}$$

$$\frac{3x}{x^2 - 16} = \frac{3x^2 + 12x}{(x-4)(x+4)^2}$$

$$\frac{(x-4)(x+4)^2}{(x+4)^2} = (x-4)$$

$$(x-4) \cdot 4x = 4x(x-4), \text{ or } 4x^2 - 16x$$

$$\frac{4}{x^2 + 8x + 16} = \frac{4x^2 - 16x}{(x-4)(x+4)^2}$$

$$\frac{3x^2 + 12x}{(x-4)(x+4)^2} + \frac{4x^2 - 16x}{(x-4)(x+4)^2} = \frac{7x^2 - 4x}{(x-4)(x+4)^2}$$

Practice Questions

Express each of the following in simplest form:

1 $\dfrac{8pq^2}{4p^2q}$

2 $\dfrac{3m+9}{5m+15}$

3 $\dfrac{x^2-6x+9}{x^2-9}$

For each of the following, perform the indicated operation and express the answer in simplest form.

4 $\dfrac{4x^3}{9y^2}\cdot\dfrac{3y}{6x}$

5 $\dfrac{y-2}{2y+2}\cdot\dfrac{5y+5}{4y-8}$

6 $\dfrac{8y^3(y^2-4)}{15(x-4)}\cdot\dfrac{5x(x^2-8x+16)}{4y(y-2)(y^2+5y+6)}$

7 $\dfrac{6a+3}{4}\div\dfrac{2a+1}{8}$

8 $\dfrac{(e-1)(f+2)g}{(r-1)(s+2)t}\div\dfrac{(e-1)^2g^4}{(r-1)(s+2)}$

9 $\dfrac{36(m+2)}{m^5(m-2)^4}\div\dfrac{60(m+2)(m-2)}{10}$

10 $\dfrac{5y}{4xy^2}+\dfrac{6y}{x^2y^2}$

11 $\dfrac{4c}{5ab^2}+\dfrac{7}{a^2b}$

12 $\dfrac{8}{3r^2}-\dfrac{y}{5r}$

13 $\dfrac{9}{3rs}-\dfrac{4}{2r^2s^3}$

Chapter Practice Answers

1 $\dfrac{8pq^2}{4p^2q} = \dfrac{\overset{2}{\cancel{8}}p\cancel{q}^{\cancel{2}}}{\underset{1}{\cancel{4}}p^{\cancel{2}}\cancel{q}} = \dfrac{2q}{p}$

2 $\dfrac{3m+9}{5m+15} = \dfrac{3(m+3)}{5(m+3)}$

$\dfrac{3\cancel{(m+3)}}{5\cancel{(m+3)}} = \dfrac{3}{5}$

3 $\dfrac{x^2-6x+9}{x^2-9} = \dfrac{(x-3)^2}{(x-3)(x+3)}$

$\dfrac{(x-3)^{\cancel{2}}}{\cancel{(x-3)}(x+3)} = \dfrac{x-3}{x+3}$

4 $\dfrac{\overset{2}{\cancel{4}}x^3}{\underset{3}{\cancel{9}}y^2} \cdot \dfrac{\overset{1}{\cancel{3}}x}{\underset{3}{\cancel{6}}x} = \dfrac{2x^{\cancel{3}2}\cancel{x}}{9\cancel{x}y^{\cancel{2}1}} = \dfrac{2x^2}{9y}$

5 $\dfrac{y-2}{2y+2} \cdot \dfrac{5y+5}{4y-8} = \dfrac{y-2}{2(y+1)} \cdot \dfrac{5(y+1)}{4(y-2)}$

$\dfrac{\cancel{y-2}^{1}}{2\cancel{(y+1)}} \cdot \dfrac{5\cancel{(y+1)}}{4\cancel{(y-2)}} = \dfrac{5}{8}$

6 $\dfrac{8y^3(y^2-4)}{15(x-4)} \cdot \dfrac{5x(x^2-8x+16)}{4y(y-2)(y^2+5y+6)} = \dfrac{\overset{2}{\cancel{8}}y^{\cancel{3}2}\cancel{(y-2)}(y+2)}{\underset{3}{\cancel{15}}\cancel{(x-4)}} \cdot \dfrac{\cancel{5}x(x-4)^{\cancel{2}1}}{4y\cancel{(y-2)}\cancel{(y+2)}(y+3)}$

$\qquad\qquad = \dfrac{2y^2}{3} \cdot \dfrac{x(x-4)}{(y+3)}$

$\qquad\qquad = \dfrac{2xy^2(x-4)}{3(y+3)}$

7 $\dfrac{6a+3}{4} \div \dfrac{2a+1}{8} = \dfrac{6a+3}{4} \cdot \dfrac{8}{2a+1}$

$\dfrac{3\cancel{(2a+1)}}{\underset{1}{\cancel{4}}} \cdot \dfrac{\cancel{8}^{2}}{\underset{1}{\cancel{(2a+1)}}} = \dfrac{6}{1} = 6$

8 $\dfrac{(e-1)(f+2)g}{(r-1)(s+2)t} \div \dfrac{(e-1)^2 g^4}{(r-1)(s+2)} = \dfrac{(e-1)(f+2)g}{(r-1)(s+2)t} \cdot \dfrac{(r-1)(s+2)}{(e-1)^2 g^4}$

$\dfrac{\cancel{(e-1)}(f+2)\cancel{g}}{\cancel{(r-1)(s+2)}t} \cdot \dfrac{\cancel{(r-1)(s+2)}}{(e-1)^{\cancel{2}1} g^{\cancel{4}3}} = \dfrac{(f+2)}{g^3 t (e-1)}$

9 $\dfrac{36(m+2)}{m^5(m-2)^4} \div \dfrac{60(m+2)(m-2)}{10} = \dfrac{36(m+2)}{m^5(m-2)^4} \cdot \dfrac{10}{60(m+2)(m-2)}$

$\dfrac{\cancel{36}(m+2)}{m^5(m-2)^4} \cdot \dfrac{\cancel{10}}{\cancel{60}(m+2)(m-2)} = \dfrac{6}{m^5(m-2)^5}$

10 $\dfrac{5y}{4xy^2} + \dfrac{6y}{x^2 y^2} = \dfrac{5xy}{4x^2 y^2} + \dfrac{4 \cdot 6y}{4x^2 y^2}$

$\dfrac{5xy}{4x^2 y^2} + \dfrac{4 \cdot 6y}{4x^2 y^2} = \dfrac{5xy + 24y}{4x^2 y^2}$

11 $\dfrac{4c}{5ab^2} + \dfrac{7}{a^2 b} = \dfrac{4ac}{5a^2 b^2} + \dfrac{7 \cdot 5b}{5a^2 b^2}$

$\dfrac{4ac}{5a^2 b^2} + \dfrac{35b}{5a^2 b^2} = \dfrac{4ac + 35b}{5a^2 b^2}$

12 $\dfrac{8}{3r^2} - \dfrac{y}{5r} = \dfrac{8 \cdot 5}{15r^2} - \dfrac{3 \cdot ry}{15r^2}$

$\dfrac{40}{15r^2} - \dfrac{3ry}{15r^2} = \dfrac{40 - 3ry}{15r^2}$

13 $\dfrac{9}{3rs} - \dfrac{4}{2r^2 s^3} = \dfrac{9 \cdot 2rs^2}{6r^2 s^3} - \dfrac{4 \cdot 3}{6r^2 s^3}$

$\dfrac{18rs^2}{6r^2 s^3} - \dfrac{12}{6r^2 s^3} = \dfrac{18rs^2 - 12}{6r^2 s^3}$

$\dfrac{\cancel{18}^3 rs^2 - \cancel{12}}{\cancel{6}_2 r^2 s^3} = \dfrac{3rs^2 - 2}{6r^2 s^3}$

chapter 11

Roots and Radicals

The subject of roots and radicals was introduced in Chapter 1, "The Basics." *Square roots* have nothing to do with square trees. In fact, they are the opposites of numbers that are raised to the second power. When you raise a number to the second power, you'll recall, you multiply it by itself. To find the square root of a number, find the number which, when multiplied by itself, makes that number.

$2 \times 2 = 4$, so 2 is the square root of 4. Although square roots are not the only type of root in algebra, it is by far the most often used. You can indicate square root in algebra in several ways, but the most common is probably by using the *radical sign*. I use red here, because in the twentieth century it was the color most often used to represent radicalism. Of course, radicalism has nothing to do with algebraic radicals, but I know a good thing when I see one.

As previously noted, you dealt with this subject perfunctorily back in Chapter 1, but in this chapter, you are going to take a much more in-depth look. The symbol $\sqrt{}$ is called a **radical sign**. It is used to designate the root of whatever number is beneath it. For example, $\sqrt{4}$ is read "**the square root of 4.**" Since 4 can be broken into its components like this, $\sqrt{4} = \sqrt{2 \cdot 2}$, the square root of 4 is 2. You can write that exclusively with symbols as follows: $\sqrt{4} = 2$.

PERFECT SQUARES

When a number has a square root that is a whole number, that number is said to be a **perfect square**. The total size of the realm of perfect squares is infinite, but those equal to 100 or less are 1, 4, 9, 16, 25, 36, 49, 64, 81, and 100. Can you see the pattern? If you can't, start by figuring out what the square root of each of those listed numbers is. But it doesn't stop at squares and square roots. A number is called a **square** of another based on the geometric figure of the same name.

Look at the top right figure, and you'll see what I mean.

Notice that each of the edges of the square is the square root. The area is the actual square.

Give the square a third dimension by adding depth. Check the second figure out:

Here, each edge is drawn in purple, while the volume of the figure is dark green. Each edge is the "cube root" of the figure's volume: $l = \sqrt[3]{V}$.

CUBE ROOTS AND CUBIC NUMBERS

Notice the little 3 on the radical sign. That tells you it is a **cube root** that is being represented. Suppose the volume of the cube is 8 cubic units. Then to find each edge you solve the following:

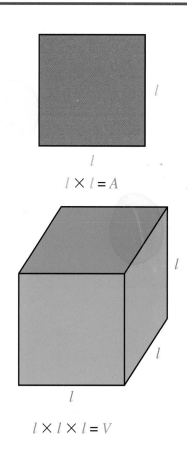

l

l

$l \times l = A$

l

l

l

l

$l \times l \times l = V$

$$\sqrt[3]{8} = \sqrt[3]{2 \cdot 2 \cdot 2}$$
$$\sqrt[3]{8} = 2$$

The first cubic number is 1. It is equal to $1 \cdot 1 \cdot 1$, so you can properly write that $\sqrt[3]{1} = 1$. The next cubic number is 8; then 27, then 64, and 125. Notice that cubic numbers get large very quickly. **Cubic numbers**, unlike square numbers, have one and only one cubic root. Each square number has exactly two square roots. "How can that be?" you ask. Well consider the following: What's the value of 2^2? "That's an easy one," you're thinking; "why, it's 4." If that's what you were thinking, you're absolutely right, but in that case, what's the value of $(-2)^2$? Whoa! That's also 4. Now are you getting the idea: *For every positive square root of a number, there's also a negative one.* You could say that for any perfect square, it's the product of either the positive or negative value of its square root—that is, $\sqrt{x^2} = \pm x$. There is one additional unique property of square roots—well, make that of even-numbered roots. Negative integers don't have any—at least not in the realm of real numbers. The square root of –4, for example, is *imaginary*. The variable i is an imaginary value that stands for $\sqrt{-1}$, so the square root of –4 would be represented as $2i$. You won't look any further at imaginary numbers in this book, since it's really a topic for Algebra II.

Perfect cubes can be positive or negative. The cube root of 8 is 2, and the cube root of –8 is –2. Try it. $-2 \times -2 = 4 \times -2 = -8$.

No convenient geometric representations exist for numbers raised to powers greater than three, but that does not mean there are no roots greater than cube roots.

$$2 \times 2 \times 2 \times 2 = 16$$
$$\sqrt[4]{16} = 2$$

That is read "the fourth root of 16 is 2."

Read that "the fifth root of 243 is 3.

$$3 \times 3 \times 3 \times 3 \times 3 = 243$$
$$\sqrt[5]{243} = 3$$

TIP

There is really no limit to the "th" of the root. That limitless number is sometimes referred to as the nth root of a number. Those numbers that have odd-numbered roots may be positive or negative; those with even-numbered roots may only be positive.

Operations with Square Roots

It is possible to perform many different operations with square roots. Some of them can be performed under a single radical sign, while others may involve two or more radical signs. In many respects the rules for operating with square roots follow the general rules for operating with variables. It's important to pay careful attention to the special rules governing the various operations with square roots.

UNDER A SINGLE RADICAL SIGN

Notice you multiplied the two factors under the sign together to get 36 and then found the square root of that, which is 6.

$$\sqrt{(4)(9)} = \sqrt{36} = 6$$

But you could have solved it a different way:

First you found the square root of 4, 2. Then you removed that 2 from under the radical sign. Next, you found the square root of 9, 3, and removed it from under the radical sign. Note that when something is removed from under the radical it is multiplied by whatever was already there. Finally, you find that 2 • 3 = 6. You wound up at the same place, but it was a bit more circuitous a route.

$$\sqrt{(4)(9)} = \sqrt{2 \cdot 2\,(9)}$$
$$2\sqrt{9} = 2\sqrt{3 \cdot 3}$$
$$2 \cdot 3 = 6$$

Next, consider addition and subtraction under a single radical sign:

Adding 40 + 24 under the radical sign gave you the square root of 64. On the second line you took a different approach from what you did in the multiplication, rewriting the square root of 64 as the square root of 8^2.

$$\sqrt{40 + 24} = \sqrt{64}$$
$$\sqrt{64} = \sqrt{8^2} = 8$$

The purpose here is to emphasize that finding a square root is the same as undoing an exponent of 2.

Here you have followed the same pattern in demonstrating subtraction of 5 from 54 under the radical sign. This time it was 7^2.

$$\sqrt{54 - 5} = \sqrt{49}$$
$$\sqrt{49} = \sqrt{7^2} = \sqrt{7}$$

Operations with Square Roots *(continued)*

Division is mostly the same here, but with a little twist:

$$\sqrt{\frac{75}{3}} = \sqrt{25}$$
$$\sqrt{25} = \sqrt{5^2} = 5$$

Note: $\sqrt{\frac{75}{3}} = \frac{\sqrt{75}}{\sqrt{3}}$

The note, above, is how division of radicals differs from addition and subtraction, but not from multiplication. That is:

$$\sqrt{2} \cdot \sqrt{4} = \sqrt{8}$$

Last but not least, consider this:

$$\sqrt{9+4} = \sqrt{13}$$

Think about that, because at first blush you probably noticed that both 9 and 4 are perfect squares, yet they add together to make a number with no exact square root.

WITH LIKE RADICALS

Addition and subtraction of radicals is possible only when the numbers under the radical signs are the same. As noted earlier, in many respects the rules for operating with square roots follow the general rules for operating with variables. This is especially true for the operations of addition and subtraction. Consider the fact that $2x + 3x = 5x$ and $6y - y = 5y$.

When adding or subtracting radicals, the radicals behave as if they were the variables, so:

$$3\sqrt{2} + 2\sqrt{2} = 5\sqrt{2}$$

The numerical coefficients are added as with the variable addition above. The $\sqrt{2}$ just names what it is that's being added; like apples, oranges, or x's.

For subtraction:

$$6\sqrt{3} - \sqrt{3} = 5\sqrt{3}$$

Note that just as is the case with variables, a radical with no numerical coefficient before it is assumed to have a 1 in front of it, so $\sqrt{3} = 1\sqrt{3}$, and is treated as such when adding or subtracting.

Consider the following addition and subtraction:

$$3\sqrt{3} + 4\sqrt{5} = 3\sqrt{3} + 4\sqrt{5}$$
$$\text{and}$$
$$6\sqrt{5} - 2\sqrt{7} = 6\sqrt{5} - 2\sqrt{7}$$

Again, they behave in a manner identical to variables. If the variables are not identical, you can't add them; if the radicals are not identical, you can't add them. And if the variables are not identical, you can't subtract them; if the radicals are not identical, you can't subtract them.

Always be careful not to fall into temptations like these:

$$\sqrt{30} + \sqrt{6} \neq \sqrt{36}$$
$$\sqrt{54} - \sqrt{5} \neq \sqrt{49}$$

TIP

Remember, what's under the radical signs must be identical or they cannot be added or subtracted.

ADDITION AND SUBTRACTION OF SIMPLIFIED ROOTS

Sometimes radicals do not look like they can be added or subtracted, but they are hiding their true identities, sort of like radical wolves in radical sheep's clothing, for example:

$$4\sqrt{3} + 2\sqrt{27}$$

At first, these look like two incompatible radicals, until you take a closer look at the second one. That second radical is not in its simplest form. Factor 27, and you'll get 3×9:

$$4\sqrt{3} + 2\sqrt{27}$$
$$2\sqrt{27} = 2\sqrt{9 \cdot 3} = 2\sqrt{3^2 \cdot 3} = 2 \cdot 3\sqrt{3} = 6\sqrt{3}$$
$$4\sqrt{3} + 6\sqrt{3} = 10\sqrt{3}$$

Did you follow all of that? The main thing is that

$$2\sqrt{27} = 6\sqrt{3}$$
$$\text{so that } 4\sqrt{3} + 6\sqrt{3} = 10\sqrt{3}$$

How about this one?

You'll get some more practice with these later in the "Chapter Practice" section. The key piece of learning to take away from this lesson is that just because two radicals don't seem identical at first blush, don't dismiss them as different until you've thoroughly analyzed the situation.

$$4\sqrt{50} - 7\sqrt{2}$$
$$4\sqrt{50} = 4\sqrt{2 \cdot 25} = 4\sqrt{2 \cdot 5^2} = 4 \cdot 5\sqrt{2} = 20\sqrt{2}$$

$$\text{Therefore, } 20\sqrt{2} - 7\sqrt{2} = 13\sqrt{2}$$

MULTIPLYING NON-NEGATIVE ROOTS

When multiplying roots that are constants, you know that negative square roots are not allowed, so you simply proceed, as in:

$$\sqrt{8} \cdot \sqrt{2} = \sqrt{16} = 4$$

When dealing with radicals containing variables, however, it is essential to realize that the non-negative roots are the only ones involved in the product.

$$\sqrt{a^3} \cdot \sqrt{a^7} = \sqrt{a^{3+7}} = \sqrt{a^{10}}$$

Did you remember that when both factors being multiplied are the same variable, the exponents are added? That's why you add $3 + 7$ to get 10. But wait, you're not done with this yet, since you're looking at an even exponent under a square root sign:

$$\sqrt{a^{10}} = \sqrt{\left(a^5\right)^2} = a^5$$

So it all comes down to a^5.

Try this one:

In the first line, the 6 and 3 were multiplied together, as were the x's. On the second line, the 18 and 2 are multiplied together, and the x^2 was multiplied by the y^4. Finally, since each member is a perfect square in its own right, you get the final answer: $6xy^2$. Answers should always be expressed in the simplest possible form.

$$\sqrt{6x} \cdot \sqrt{3x} \cdot \sqrt{2y^4} = \sqrt{18x^2} \cdot \sqrt{2y^4}$$
$$\sqrt{18x^2} \cdot \sqrt{2y^4} = \sqrt{2 \cdot 18x^2 y^4} = \sqrt{36x^2 y^4}$$
$$\sqrt{36x^2 y^4} = 6xy^2$$

Here's one that's slightly different, although the principles are exactly the same:

$$\sqrt{a^2 b} \cdot \sqrt{bc^2} \cdot \sqrt{acd} \cdot \sqrt{bd} = \sqrt{a^2 b^2 c^2} \cdot \sqrt{abcd^2}$$

First, combine the first two radicals and the last two radicals. Then combine what's left:

$$\sqrt{a^2 b^2 c^2} \cdot \sqrt{abcd^2} = \sqrt{a^3 b^3 c^3 d^2}$$

Now break that down. Bear in mind that like variables, radicals don't need to be separated by multiplication signs to indicate multiplication.

Putting each variable under its own radical sign makes it easier to see which part of what's under it can be brought out. Then, the final step is to put it all back together.

$$\sqrt{a^3 b^3 c^3 d^2} = \sqrt{a^3} \sqrt{b^3} \sqrt{c^3} \sqrt{d^2}$$
$$\sqrt{a^3} \sqrt{b^3} \sqrt{c^3} \sqrt{d^2} = d\sqrt{a \cdot a^2} \sqrt{b \cdot b^2} \sqrt{c \cdot c^2}$$
$$= ad\sqrt{a} \cdot b\sqrt{b} \cdot c\sqrt{c}$$
$$= abcd\sqrt{abc}$$

DIVIDING NON-NEGATIVE ROOTS

For the purposes of dividing roots, all roots are assumed to be non-negative. For any division of radicals, the following is true:

$$\sqrt{\frac{a}{b}} = \frac{\sqrt{a}}{\sqrt{b}}$$

Any fraction with a radical sign in its denominator is by definition irrational. It is traditional and accepted practice to never leave any fractional answer as irrational. To rationalize a denominator, multiply the fraction by 1, in the form of its conjugate root over itself. In the case of a number whose denominator is a pure radical, such as $\frac{\sqrt{a}}{\sqrt{b}}$, the conjugate root would be $\frac{\sqrt{b}}{\sqrt{b}}$. To rationalize the denominator, multiply by that over itself (which is equal to 1), as follows: $\frac{\sqrt{a}}{\sqrt{b}} \cdot \frac{\sqrt{b}}{\sqrt{b}} = \frac{\sqrt{ab}}{b}$. Ta-da! That was easy enough, wasn't it? If the denominator were in the form $a + \sqrt{b}$, then the conjugate root would be $a - \sqrt{b}$.

In order to rationalize the denominator of such a fraction without changing its value, you would proceed as follows:

$$\frac{c}{a + \sqrt{b}} \cdot \frac{a - \sqrt{b}}{a - \sqrt{b}} = \frac{ac - c\sqrt{b}}{a^2 - b}$$

You should recognize the resulting denominator as resulting in the difference of two squares, even though a square root was one of the contributors.

When the denominator is of the form $a - \sqrt{b}$, the conjugate would be $a + \sqrt{b}$, and I think we can both do without a demonstration.

For this division, solve and leave the answer with a rational denominator (where appropriate).

$$\frac{3\sqrt{5}}{2\sqrt{24}}$$

The first thing to do is to simplify the denominator:

$$\frac{3\sqrt{5}}{2\sqrt{24}} = \frac{3\sqrt{5}}{2\sqrt{4\cdot6}} = \frac{3\sqrt{5}}{2\sqrt{2^2\cdot6}} = \frac{3\sqrt{5}}{2\cdot2\sqrt{6}} = \frac{3\sqrt{5}}{4\sqrt{6}}$$

Now it's time to rationalize the denominator:

$$\frac{3\sqrt{5}}{4\sqrt{6}}\cdot\frac{\sqrt{6}}{\sqrt{6}} = \frac{3^{1}\sqrt{30}}{24_{8}} = \frac{\sqrt{30}}{8}$$

And that's as far as you can go. You could also express that as $\frac{1}{8}\sqrt{30}$.

For this division, solve and leave the answer with a rational denominator (where appropriate).

$$\frac{8\sqrt{3y}}{\sqrt{36y^5}}$$

Here, it is beneficial to combine both y radicals into a single fraction:

$$\frac{8\sqrt{3y}}{\sqrt{36y^5}} = 8\sqrt{\frac{3y}{36y^5}} = 8\sqrt{\frac{1}{12y^4}}$$

Now you'll revert to one radical over the other. Remember, the 8 goes with the top one:

$$\frac{8\sqrt{1}}{\sqrt{12y^4}} = \frac{8\cdot1}{\sqrt{3\cdot2^2\left(y^2\right)^2}} = \frac{8}{2y^2\sqrt{3}}$$

Finally, rationalize the denominator:

$$\frac{8}{2y^2\sqrt{3}}\cdot\frac{\sqrt{3}}{\sqrt{3}} = \frac{8^{4}\sqrt{3}}{2y^2\cdot3_{1}} = \frac{4\sqrt{3}}{3y^2}$$

Practice Questions

Answer each question to the best of your ability.

1 How is 5 related to 125 in a sense that relates to this chapter?

2 Find $\sqrt{169}$.

3 Name all square roots of 121.

4 Find the cube root of –27.

5 Find all the roots of $\sqrt[5]{1024}$.

Solve each of the following. Express each answer in simplest possible but complete terms. If the solution contains an irrational denominator, rationalize it.

6 $\sqrt{(5)(20)}$

7 $\sqrt{110-29}$

8 $\sqrt{35+14}$

9 $\sqrt{(15)(5)}$

10 $7\sqrt{2}-\sqrt{98}$

11 $5\sqrt{18}+2\sqrt{32}$

12 $3\sqrt{72}-4\sqrt{32}$

13 $\sqrt{4a^2}\cdot\sqrt{8a^2}\cdot\sqrt{2b^4}$

14 Rationalize: $\dfrac{2\sqrt{3n}}{5-\sqrt{2n}}$

15 $\sqrt{w^2x}\cdot\sqrt{xy^2}\cdot\sqrt{w^2y^2z}\cdot\sqrt{xz}$

16 $\sqrt{10x^3}\cdot\sqrt{5xy^2}\cdot\sqrt{2y^2}$

17 $\dfrac{5x}{2-\sqrt{3y}}$

18 $\dfrac{1+\sqrt{3}}{5-\sqrt{3}}$

Chapter Practice Answers

1 5 is the cube root of 125, or $5 = \sqrt[3]{125}$

2 $\sqrt{169} = \pm 13$ because $13 \cdot 13 = 169$, and $-13 \cdot -13 = 169$

3 11 and -11

4 -3

5 $\sqrt[5]{1024}$ and only 4

6 $\sqrt{(5)(20)} = \sqrt{100} = \pm 10$

7 $\sqrt{110 - 29} = \sqrt{81} = \pm 9$

8 $\sqrt{35 + 14} = \sqrt{49} = \pm 7$

9 $\sqrt{(15)(5)} = \sqrt{75} = \sqrt{25 \cdot 3} = \sqrt{5^2 \cdot 3} = 5\sqrt{3}$

10
$$\begin{aligned}
7\sqrt{2} - \sqrt{98} &= 7\sqrt{2} - \sqrt{49 \cdot 2} \\
&= 7\sqrt{2} - \sqrt{7^2 \cdot 2} \\
&= 7\sqrt{2} - 7\sqrt{2} \\
&= 0
\end{aligned}$$

11
$$\begin{aligned}
5\sqrt{18} + 2\sqrt{32} &= 5\sqrt{9 \cdot 2} + 2\sqrt{16 \cdot 2} \\
&= 5\sqrt{3^2 \cdot 2} + 2\sqrt{4^2 \cdot 2} \\
&= 5 \cdot 3\sqrt{2} + 2 \cdot 4\sqrt{2} \\
&= 15\sqrt{2} + 8\sqrt{2} \\
&= 23\sqrt{2}
\end{aligned}$$

12
$$\begin{aligned}
3\sqrt{72} + 4\sqrt{32} &= 3\sqrt{36 \cdot 2} + 4\sqrt{16 \cdot 2} \\
&= 3\sqrt{6^2 \cdot 2} + 4\sqrt{4^2 \cdot 2} \\
&= 3 \cdot 6\sqrt{2} + 4 \cdot 4\sqrt{2} \\
&= 18\sqrt{2} + 16\sqrt{2} \\
&= 34\sqrt{2}
\end{aligned}$$

⑬ $\sqrt{4a^2} \cdot \sqrt{8a^2} \cdot \sqrt{2b^4} = \sqrt{32a^4} \cdot \sqrt{2b^4}$

$\qquad \sqrt{32} \cdot \sqrt{2b^4} = \sqrt{2 \cdot 32a^4 b^4} = \sqrt{64a^4 b^4}$

$\qquad \qquad \sqrt{64a^4 b^4} = 8a^2 b^2$

⑭ $\dfrac{2\sqrt{3n}}{5-\sqrt{2n}} \cdot \dfrac{5+\sqrt{2n}}{5+\sqrt{2n}} = \dfrac{10\sqrt{3n}+\sqrt{6n^2}}{25-2n} = \dfrac{10\sqrt{3n}+n\sqrt{6}}{25-2n}$

⑮ $\sqrt{w^2 x} \cdot \sqrt{xy^2} \cdot \sqrt{w^2 y^2 z} \cdot \sqrt{xz} = \sqrt{w^2 x^2 y^2} \cdot \sqrt{w^2 xy^2 z^2}$

$\qquad \sqrt{w^2 x^2 y^2} \cdot \sqrt{w^2 xy^2 z^2} = \sqrt{w^4 x^3 y^4 z^2}$

$\qquad \qquad \qquad = \sqrt{\left(w^2\right)^2 \left(x^2\right) x \left(y^2\right)^2 z^2}$

$\qquad \qquad \qquad = w^2 xy^2 z \sqrt{x}$

⑯ $\sqrt{10x^3} \cdot \sqrt{5xy^2} \cdot \sqrt{2y^2} = \sqrt{50x^4 y^2} \cdot \sqrt{2y^2}$

$\qquad \qquad \qquad = \sqrt{100x^4 y^4}$

$\qquad \qquad \qquad = \sqrt{(10)^2 \left(x^2\right)^2 \left(y^2\right)^2}$

$\qquad \qquad \qquad = 10x^2 y^2$

⑰ $\dfrac{5x}{2-\sqrt{3y}} \cdot \dfrac{2+\sqrt{3y}}{2+\sqrt{3y}} = \dfrac{10x+5x\sqrt{3y}}{4-3y}$

⑱ $\dfrac{1+\sqrt{3}}{5-\sqrt{3}} \cdot \dfrac{5+\sqrt{3}}{5+\sqrt{3}} = \dfrac{5+6\sqrt{3}+3}{25-3}$

$\qquad \qquad = \dfrac{8+6\sqrt{3}}{22}$

$\qquad \qquad = \dfrac{4+3\sqrt{3}}{11}$

Quadratic Equations

O kay, this is the part you've been waiting for—
you know, the hard stuff. (Just kidding.) Too
many people have had a difficult time with
quadratic equations because they were not properly
prepared to deal with them. If you've understood all
the material that has been covered so far, you should
have no trouble here. **Quadratic equations** contain
a term raised to the exponent ^{2}. Each equation has
either two solutions (a positive one and a negative
one), one solution, or no solution. If it has no
solution, that means that there are no real numbers
that conform to the requirements of the equation.
There are three different ways to solve quadratic
equations, and you shall study all three of them
here—one at a time.

A quadratic equation is one that can be written in the form:

$$ax^2 + bx + c = 0$$

The a, b, and c are normally replaced by constants, but they are used to signify first, second, and third positions in the equation when it is in its standard form. The zero on the right side of the equal sign is no accident either. It is part of the requirement for the equation's being in standard quadratic form.

No quadratic equation ever contains a variable raised to a power higher than two. Also, there must always be a variable raised to the exponent^2, or it's not a quadratic equation. The $+$ signs are a little more flexible. Since a constant may be negative as well as positive, any of the following may occur and still be in standard quadratic form:

$$ax^2 + bx - c = 0$$
$$ax^2 - bx + c = 0$$
$$ax^2 - bx - c = 0$$

You're now going to look at a few quadratic equations that are *not* in standard form. The objective will be to get every one of them into standard form. Start out with this one:

$$5x + 4x^2 = 21$$

The two problems with this equation are that the a and b terms are reversed, and the c term is on the wrong side of the equal sign. Reversing the first two is easy enough, and to finish the job you'll subtract 21 from each side.

$$4x^2 + 5x - 21 = 21 - 21$$
$$4x^2 + 5x - 21 = 0$$

Now look at the next quadratic:

$$x^2 - 9 = 0$$

Where are the a and b terms? Well, the a term is right where it belongs, but the b term is missing. To put it into standard form you need to add a b term that won't change the value of the equation. Where's that a, you ask? All right, I'll put it there for you, but you're not going to like it.

$$1x^2 + 0x - 9 = 0$$

It's $a = 1$, which you never write, but which I did this time to tease you; adding a $0x$ doesn't change the value of the equation, since it equals 0. The fact of the matter is, when solving quadratic equations by factoring (next lesson), you love to see the minus sign and the missing b term. You should recognize that as the difference of two squares, which you studied earlier when you practiced factoring.

Finally, consider this equation:

$$3x = 1 + 2x^2$$

This equation presents a couple of interesting choices. You could subtract $1 + 2x^2$ from both sides, and then rearrange the terms. That would give you the equation:

$$3x - (1 + 2x^2) = 1 + 2x^2 - (1 + 2x^2)$$
$$-2x^2 + 3x - 1 = 0$$

Or you could rearrange the terms on the right and subtract $3x$ from both sides, like this:

$$3x = 1 + 2x^2$$
$$3x - 3x = 2x^2 + 1 - 3x$$
$$0 = 2x^2 - 3x + 1$$

Now, since an equation is like an equal arm balance (remember Chapter 4), if it were a balance you'd just switch pans. In this case, you pivot on the equal sign.

$$2x^2 - 3x + 1 = 0$$

You would have gotten the same result if you had multiplied the first equation you found by –1. Remember, you can multiply an equation by anything without changing its value, *as long as you do it to every term.*

Solving Quadratic Equations by Factoring

When you studied factoring in Chapter 7, you might have thought it was just for fun, and it was at the time. Now, however, you're going to get the chance to put it to some use. If you don't remember the various forms of factoring you did in Chapter 7, go back and review it. I'll wait. Not all quadratic equations can be solved by factoring. Indeed, most of them cannot. But when the opportunity is there, it should be used.

The first thing that needs to be done is to put the equation into standard form. I'm not going to waste your time by bothering you with that now. You already saw it in the last lesson, so start off with equations that are already in standard form, starting with this:

$$y^2 + 2y + 1 = 0$$

Since you went back and reviewed Chapter 7 if you felt the need to, I'm not going to waste time or space analyzing this equation. It factors as follows:

$$(y + 1)(y + 1) = 0$$

Next, set each of those factors equal to 0 and solve separately.

$$y + 1 = 0 \qquad y + 1 = 0$$
$$y = -1 \qquad y = -1$$

Not surprisingly, the answers are the same, so the roots of this equation are $(-1, -1)$. Pretty cool eh?

Try another one:

$$x^2 - 9 = 0$$

Does that look strangely familiar? It's the difference-of-two-squares equation that you played with in the last lesson, and it is the exception to the "get it into standard form" rule. You do remember the difference of two squares, I hope. Go for it:

$$x^2 - 9 = 0$$
$$(x + 3)(x - 3) = 0$$
$$x + 3 = 0 \qquad x - 3 = 0$$
$$x = -3 \qquad x = 3$$

Do you know why I love writing math books? It's because it's so much easier to solve equations that you've made up yourself. But you probably didn't want to know that. Try this one:

$$2z^2 - 5z - 3 = 0$$

This one isn't quite as easy to factor as the two previous ones, but it's not too bad. The two minus signs should help steer you in the right direction, as should 2, 3, and 5.

$$2z^2 - 5z - 3 = 0$$
$$(2z + 1)(z - 3) = 0$$
$$2z + 1 = 0 \qquad z - 3 = 0$$
$$2z = -1 \qquad z = 3$$
$$z = -\frac{1}{2}$$

It's always a good idea to check your answer by substituting the root(s) that you found back into the original equation. If the root is correct, the solution of the equation will yield a true statement. Here, I've worked out the last solution, substituting both roots into the original equation.

$$2\left(-\frac{1}{2}\right)^2 - 5\left(-\frac{1}{2}\right) - 3 = 0 \qquad \text{and/or} \qquad 2(3)^2 - 5(3) - 3 = 0$$
$$2\left(\frac{1}{4}\right) + \frac{5}{2} - 3 = 0 \qquad\qquad 2(9) - 15 - 3 = 0$$
$$\frac{2}{4} + \frac{5}{2} - 3 = 0 \qquad\qquad 18 - 15 - 3 = 0$$
$$\frac{6}{2} - 3 = 0 \qquad\qquad 18 - 18 = 0$$
$$3 - 3 = 0$$

So both roots checked out. You should go back and check out the answers to the two previous equations.

$$x^2 - 7x = 0$$
$$x(x - 7) = 0$$
$$x = 0 \qquad x - 7 = 0$$
$$x = 7$$

You've already seen an example of a partial quadratic equation when you worked with the difference of two squares. Here's another example of a partial quadratic that can be solved by factoring:

Pretty tricky, don't you think? Go ahead and check the two answers. You'll see that both work.

Solving Quadratics by Completing the Square

A second method for solving quadratic equations is known as completing the square. **Obviously, it is most often used for equations that are not readily factorable. There are several steps to be followed in completing the square, but the method can be used whether the roots are real or imaginary. The purpose is to create a perfect square on the left side of the equal sign that can readily be factored to form a square of the form $(x + n)^2 = y$. At that point, $x + n$ can be set equal to $\pm y$, and solved.**

The steps for completing the square are as follows:

❶ Write the equation in the form $ax^2 + bx = -c$. This time there are no exceptions.

❷ a must equal 1. If it does not, divide the entire equation by a or multiply the equation by $\frac{1}{a}$ so that $a = 1$. Otherwise, you cannot proceed.

❸ Take half of the value of b, square it, and add it to both sides of the equation. At this point, the left side of the equation will be a perfect square.

❹ Rewrite the equation as $(x + n)^2 = y$, where y is whatever's on the right of the equals sign.

❺ Finally, split the terms from step 4 to make $x + n = \sqrt{y}$ and $x + n = -\sqrt{y}$ and solve both for x, the root(s).

Try one to see how it works:

$$3x^2 - 18x + 15 = 0$$

First you'll have to subtract 15 from each side.

$$3x^2 - 18x + 15 - 15 = -15$$
$$3x^2 - 18x = -15$$

Now there's the matter of that a term. It's not 1, so you need to divide by 3.

$$\frac{3x^2 - 18x}{3} = \frac{-15}{3}$$
$$x^2 - 6x = -5$$

Next, take half of the value of b, square it, and add it to both sides.

$$x^2 - 6x + (-3)^2 = -5 + (-3)^2$$
$$x^2 - 6x + 9 = -5 + 9$$
$$x^2 - 6x + 9 = 4$$

Now it's time to factor the perfect square you just created on the left side.

$$(x - 3)^2 = 4$$

If you're not sure why that's an $x - 3$, look at the sign of the b term in the equation three lines up. Finally, split both terms and solve each for the positive and negative values of the square root of the right side.

$x - 3 = \sqrt{4}$	and/or	$x - 3 = -\sqrt{4}$
$x - 3 = 2$		$x - 3 = -2$
$x = 3 + 2$		$x = 3 - 2$
$x = 5$		$x = 1$

Substitute each root back into the original equation to see whether it works.

That works for me!

$$3(5)2 - 18 \cdot 5 + 15 = 0$$
$$3(25) - 90 + 15 = 0$$
$$75 - 90 + 15 = 0$$
$$-15 + 15 = 0 \text{ and/or } 3(1)^2 - 18 \cdot 1 + 15 = 0$$
$$3(1) - 18 \cdot 1 + 15 = 0$$
$$3 - 18 + 15 = 0$$
$$-15 + 15 = 0$$

Try one more of those. How about:

$$\frac{3}{2}y^2 + 15y + 21 = 0$$

First you need to rewrite the equation, subtracting 21 from both sides.

$$\frac{3}{2}y^2 + 15y + 21 - 21 = -21$$
$$\frac{3}{2}y^2 + 15y = -21$$

Next, you have to do something about that $\frac{3}{2}$. The solution is to multiply everything by its reciprocal, $\frac{2}{3}$.

$$\frac{2}{3}\left(\frac{3}{2}y^2 + 15y = -21\right)$$
$$y^2 + 10y = -14$$

Now take half of the value of b, square it, and add it to both sides of the equation.

$$y^2 + 10y = -14$$
$$\left(\frac{10}{2}\right)^2 = (5)^2 = 25$$
$$y^2 + 10y + 25 = -14 + 25$$

Now it's time to factor the perfect square you just created on the left side.

$$(y + 5)^2 = 11$$

Finally, split both terms and solve each for the positive and negative values of the square root of the right side.

$$y + 5 = \sqrt{11} \qquad \text{and/or} \qquad y + 5 = -\sqrt{11}$$
$$y = -5 + \sqrt{11} \qquad\qquad\qquad y = -5 - \sqrt{11}$$

I'm going to leave it to you to check that problem, because I want to do something that you are probably going to find insane—until you understand where I am going with it. I want to work out the solution to the following equation by completing the square:

$$ax^2 + bx = -c$$

I presume that it looks familiar, so get to it. You need to divide each
term by a.

$$\frac{ax^2}{a} + \frac{bx}{a} = -\frac{c}{a}$$

$$x^2 + \frac{bx}{a} = -\frac{c}{a}$$

Next, take half the b term's coefficient, square it, and
add it to both sides.

$$\left(\frac{1}{2}\frac{b}{a}\right)^2 = \left(\frac{b}{2a}\right)^2 = \frac{b^2}{4a^2}$$

$$x^2 + \frac{bx}{a} + \frac{b^2}{4a^2} = -\frac{c}{a} + \frac{b^2}{4a^2}$$

To combine the terms on the right side of the equals
sign, you'll need to combine both fractions over the
common denominator $4a^2$.

$$x^2 + \frac{bx}{a} + \frac{b^2}{4a^2} = -\frac{c}{a} + \frac{b^2}{4a^2} = \frac{b^2 - 4ac}{4a^2}$$

That is:

$$x^2 + \frac{bx}{a} + \frac{b^2}{4a^2} = \frac{b^2 - 4ac}{4a^2}$$

Next, factor the perfect square that you've made on the left side of the equation and split the roots to
solve for their positive and negative values.

$$\left(x + \frac{b}{2a}\right)^2 = \frac{b^2 - 4ac}{4a^2}$$

$$x + \frac{b}{2a} = \sqrt{\frac{b^2 - 4ac}{4a^2}} \qquad x + \frac{b}{2a} = -\sqrt{\frac{b^2 - 4ac}{4a^2}}$$

$$x = -\frac{b}{2a} + \sqrt{\frac{b^2 - 4ac}{4a^2}} \qquad x = -\frac{b}{2a} - \sqrt{\frac{b^2 - 4ac}{4a^2}}$$

$$x = -\frac{b}{2a} + \frac{\sqrt{b^2 - 4ac}}{2a\sqrt{4a^2}} \qquad x = -\frac{b}{2a} - \frac{\sqrt{b^2 - 4ac}}{2a\sqrt{4a^2}}$$

Solving Quadratics by Completing the Square *(continued)*

Well, it should be obvious that you can combine those fractions, since the denominators are the same, and you'll end up with the same fraction on each side, except one radical will be added and the other subtracted. Why not combine those as well, to get:

$$x = \frac{-b \pm \sqrt{b^2 - 4ac}}{2a}$$

That may seem to you like a lot of work, and it was, but it was all worthwhile, as you'll see on the next page. This equation is known as the **quadratic formula** and can be used to solve any quadratic equation, factorable or otherwise, with roots that are real or imaginary.

TIP

Now I'm going to ask you to do something you're not going to like. It is essential that you commit the quadratic formula to memory. Everybody else who has taken algebra has done it, and there is no reason why you can't. I recommend the following mantra: "x equals negative b plus or minus the square root of b^2 minus $4ac$, all over $2a$." Say it to yourself while commuting to school or work, and while you're trying to go to sleep at night, until it comes to you automatically whenever you see a quadratic equation.

You've seen the quadratic formula, $x = \dfrac{-b \pm \sqrt{b^2 - 4ac}}{2a}$, and seen how it is derived by completing the square. You've been working feverishly to memorize it and are continuing to do so, if you haven't already memorized it. Now it's time to see how to use it.

Start out with this one: $\qquad\qquad\qquad\qquad\qquad\qquad\qquad\qquad\qquad 6m^2 - 2m - 20 = 0$

That means that $a = 6$, $b = -2$, and $c = -20$.

Next, write the quadratic formula, with the caveat that you're solving for m rather than x.

$$m = \frac{-b \pm \sqrt{b^2 - 4ac}}{2a}$$

Now, you're going to substitute the values, being certain to transfer their signs, when necessary. In other words, if the number has a negative sign, that sign should be transferred with the value, as in the case of c in this equation.

$$m = \frac{-(-2) \pm \sqrt{(-2)^2 - (4)(6)(-20)}}{2 \cdot 6}$$

Next, combine what you are able to combine.

$$m = \frac{2 \pm \sqrt{4 - (4)(6)(-20)}}{2 \cdot 6}$$

$$m = \frac{2 \pm \sqrt{4 - (-480)}}{12}$$

$$m = \frac{2 \pm \sqrt{484}}{12}$$

Simplifying 484, you find it's divisible by 4.

$$m = \frac{2 \pm \sqrt{4 \cdot 121}}{12} = \frac{2 \pm 2\sqrt{121}}{12}$$

You might recognize 121 as the square of 11, so:

$$m = \frac{2 \pm 2 \cdot 11}{12} = \frac{2 \pm 22}{12} = \frac{1 \pm 11}{6}$$

Now separate the two values of m and solve:

Now wasn't that fun?

$$m = \frac{1 + 11}{6} \quad \text{and/or} \quad m = \frac{1 - 11}{6}$$

$$m = \frac{12}{6} \qquad\qquad m = \frac{-10}{6}$$

$$m = 2 \qquad\qquad m = -\frac{5}{3}$$

You'll do one more solution by quadratic formula in a bit, but first consider a part of the formula. The portion under the radical sign is known as the **discriminant**. It has already been noted that there are three possibilities for the solutions to a quadratic equation. Using the discriminant, $b^2 - 4ac$ can tell you in advance what those roots' properties will be.

1 If $b^2 - 4ac < 0$, then there are no real roots.

2 If $b^2 - 4ac = 0$, then there is exactly one real root.

3 If $b^2 - 4ac > 0$, there are two real roots.

Consider the equation you just solved. If you look back, you'll see that the discriminant was equal to 484. What does that tell you? Sure enough, the roots were 2 and $-\frac{5}{3}$; two roots, both real.

You'll get more practice with the use of the discriminant in the next section.

Practice Questions

1 Using the constants d, e, and f, write a quadratic equation for the variable x in standard form.

2 Using the constants g, h, and i, write a quadratic equation for the variable z in standard form.

Solve each of the following equations by factoring.

3 $x^2 - 5x + 6 = 0$

4 $2a^2 + 2a - 4 = 0$

5 $y^2 + 6y + 9 = 0$

6 $x^2 - 25 = 0$

7 $p^2 - 8p + 16 = 0$

Solve each of the following equations by completing the square.

8 $2x^2 - 8x - 14 = 0$

9 $r^2 + 6r + 8 = 0$

10 $12 - y^2 = 8y$

Solve each of the following using the quadratic formula.

11 $6z^2 + 11z = -4$

12 $12v^2 + 11v = 15$

13 $7x^2 + 25x + 12 = 0$

14 $4x^2 + 16x + 15 = 0$

For the following three equations, do not solve. Just use the discriminant to determine the number of real roots.

15 $2x^2 + 4x + 2 = 0$

16 $3y^2 + 2y + 3 = 0$

17 $5x^2 + 5x + 1 = 0$

Chapter Practice Answers

1 $dx^2 + ex + f = 0$

2 $gz^2 + hz + i = 0$

3

$$x = 2, 3$$
$$x^2 - 5x + 6 = 0$$
$$(x - 3)(x - 2) = 0$$
$$x - 3 = 0 \text{ and/or } x - 2 = 0$$
$$x = 3$$
$$x = 2$$

4

$$a = 1, -2$$
$$2a^2 + 2a - 4 = 0$$
$$(2a + 4)(a - 1) = 0$$
$$2a + 4 = 0 \text{ and/or } a - 1 = 0$$
$$2a = -4$$
$$a = 1$$
$$a = -2$$

5

$$y = -3, -3$$
$$y^2 + 6y + 9 = 0$$
$$(y + 3)(y + 3) = 0$$
$$y + 3 = 0 \text{ and/or } y + 3 = 0$$
$$y = -3$$
$$y = -3$$

6

$$x = 5, -5$$
$$x^2 - 25 = 0$$
$$(x + 5)(x - 5) = 0$$
$$x + 5 = 0 \text{ and/or } x - 5 = 0$$
$$x = -5$$
$$x = 5$$

7

$$p = 4, 4$$
$$p^2 - 8p + 16 = 0$$
$$(p - 4)(p - 4) = 0$$
$$p - 4 = 0 \text{ and/or } p - 4 = 0$$
$$p = -4$$
$$p = 4$$

8 $x = 2 \pm \sqrt{18}$

First, the equation must be divided through by 2, and then 7 must be added to each side of the equation.

$$2x^2 - 8x - 14 = 0$$
$$x^2 - 4x - 7 = 0$$
$$x^2 - 4x = 7$$

Next, the b is halved, squared, and added to both sides.

$$\left(\frac{-4}{2}\right)^2 = 4$$
$$x^2 - 4x + 4 = 14 + 4$$

Now factor the newly made square on the left and split the roots, taking the square root of both sides.

$$x^2 - 4x + 4 = 18$$
$$x - 2 = \pm\sqrt{18}$$
$$x = 2 \pm \sqrt{18}$$

9

$$r = -2, -4$$

$$r^2 + 6r + 8 = 0$$

$$r^2 + 6r = -8$$

$$\left(\frac{6}{2}\right)^2 = 3^2 = 9$$

$$r^2 + 6r + 9 = -8 + 9$$

$$(r + 3)^2 = 1$$

$$r + 3 = 1 \qquad \text{and/or} \qquad r + 3 = -1$$

$$r = -3 + 1 \qquad\qquad\qquad r = -3 - 1$$

$$r = -2 \qquad\qquad\qquad\quad r = -4$$

10

$$12 - y^2 = 8y$$

$$12 - y^2 + y^2 = 8y + y^2$$

$$y^2 + 8y = 12$$

$$\left(\frac{8}{2}\right)^2 = 4^2 = 16$$

$$y^2 + 8y + 16 = 12 + 16$$

$$(y + 4)^2 = 28$$

$$\sqrt{(y + 4)^2} = \pm\sqrt{28}$$

$$y + 4 = \sqrt{28} \qquad \text{and/or} \qquad y + 4 = -\sqrt{28}$$

$$y = -4 + \sqrt{4 \cdot 7} \qquad\qquad\quad y = -4 - \sqrt{4 \cdot 7}$$

$$y = -4 + 2\sqrt{7} \qquad\qquad\quad\; y = -4 - 2\sqrt{7}$$

⑪

$$z = -\frac{1}{2}, \ -1\frac{1}{3}$$
$$6z^2 + 11z = -4$$
$$6z^2 + 11z + 4 = 0$$
$$z = \frac{-b \pm \sqrt{b^2 - 4ac}}{2a}$$
$$z = \frac{-11 \pm \sqrt{11^2 - 4 \cdot 6 \cdot 4}}{2 \cdot 6}$$
$$z = \frac{-11 \pm \sqrt{121 - 96}}{12} = \frac{-11 \pm \sqrt{25}}{12} = \frac{-11 \pm 5}{12}$$
$$z = \frac{-11 + 5}{12} \qquad \text{and/or} \qquad z = \frac{-11 - 5}{12}$$
$$z = \frac{-6}{12} = -\frac{1}{2} \qquad\qquad z = \frac{-16}{12} = -1\frac{1}{3}$$

⑫

$$v = \frac{3}{4}, \ -1\frac{2}{3}$$
$$12v^2 + 11v = 15$$
$$12v^2 + 11v - 15 = 0$$
$$v = \frac{-b \pm \sqrt{b^2 - 4ac}}{2a}$$
$$v = \frac{-11 \pm \sqrt{11^2 - 4 \cdot 12(-15)}}{24}$$
$$v = \frac{-11 \pm \sqrt{121 + 720}}{24} = \frac{-11 \pm \sqrt{841}}{24}$$
$$v = \frac{-11 \pm 29}{24}$$
$$v = \frac{-11 + 29}{24} \qquad v = \frac{-11 - 29}{24}$$
$$v = \frac{18}{24} \qquad\qquad v = \frac{-40}{24}$$
$$v = \frac{3}{4} \qquad\qquad v = -1\frac{2}{3}$$

13

$$x = -3, \ -\frac{4}{7}$$

$$7x^2 + 25x + 12 = 0$$

$$x = \frac{-b \pm \sqrt{b^2 - 4ac}}{2a}$$

$$x = \frac{-25 \pm \sqrt{25^2 - 4 \cdot 7 \cdot 12}}{2 \cdot 7}$$

$$x = \frac{-25 \pm \sqrt{625 - 336}}{14} = \frac{-25 \pm \sqrt{289}}{14}$$

$$x = \frac{-25 \pm 17}{14}$$

$$x = \frac{-25 + 17}{14} \quad \text{and/or} \quad x = \frac{-25 - 17}{14}$$

$$x = \frac{-8}{14} \qquad\qquad x = \frac{-42}{14}$$

$$x = -\frac{4}{7} \qquad\qquad x = -3$$

14

$$x = -1\frac{1}{2}, \ -2\frac{1}{2}$$

$$4x^2 + 16x + 15 = 0$$

$$x = \frac{-b \pm \sqrt{b^2 - 4ac}}{2a}$$

$$x = \frac{-16 \pm \sqrt{16^2 - 4 \cdot 4 \cdot 15}}{2 \cdot 4}$$

$$x = \frac{-16 \pm \sqrt{256 - 240}}{8}$$

$$x = \frac{-16 \pm \sqrt{16}}{8}$$

$$x = \frac{-16 + 4}{8} \quad \text{and/or} \quad x = \frac{-16 - 4}{8}$$

$$x = \frac{-12}{8} \qquad\qquad x = \frac{-20}{8}$$

$$x = -1\frac{1}{2} \qquad\qquad x = -2\frac{1}{2}$$

⑮ One real root:

$$2x^2 + 4x + 2 = 0$$
$$b^2 - 4ac = 16 - 4 \cdot 2 \cdot 2$$
$$b^2 - 4ac = 16 - 16$$
$$b^2 - 4ac = 0$$

When the discriminant = 0, there is one real root.

⑯ No real root:

$$3y^2 + 2y + 3 = 0$$
$$b^2 - 4ac = 4 - 4 \cdot 3 \cdot 2$$
$$b^2 - 4ac = 4 - 36$$
$$b^2 - 4ac = -36$$

When the discriminant < 0, there is no real root.

⑰ Two real roots:

$$5x^2 + 5x + 1 = 0$$
$$b^2 - 4ac = 25 - 4 \cdot 5 \cdot 1$$
$$b^2 - 4ac = 25 - 20$$
$$b^2 - 4ac = 5$$

When the discriminant > 0, there are two real roots.

chapter 13

Algebraic Word Problems

More than for any other reason, algebra exists to solve problems. More students of algebra have difficulty with word problems, however, than with any other aspect of algebra. Figuring out what solution you are supposed to be finding, selecting the information you need, and finally creating an equation that leads you to the proper answer can be difficult. In this chapter, I'll try to help you to dissect word problems so that what is being asked for becomes apparent. Sometimes too much information is provided, and you'll need to distinguish between what is relevant and what is not. I'll also try to help you to recognize words and phrases that indicate which algebraic or arithmetic operations are to be used. Finally, you'll practice many different types of word problems. Pay close attention to the words, and they shouldn't cause you any problems.

Word problems, also called story problems, are the most practical application of algebraic skills. In your everyday life, whether at home or work, problems arise that need to be thought through and acted upon. It is very unlikely that you'll ever have a practical use for a quadratic equation solution—unless you're a physicist or a chemist—but you may encounter word problems every day.

How to Tackle Algebraic Word Problems

- The first trick to solving an algebraic word problem is to read the problem through once, just to get an idea of what it's about. Then go back and analyze the problem to determine what information you're being asked to find. Some teachers suggest that you circle that information on the page. Let your variable equal whatever it is you're seeking to find. Remember that the variable represents a number rather than a person, place, or thing.

- Often, drawing a picture will help—not a masterpiece like the Mona Lisa, but a pencil sketch with stick figures. A chart or table of values may also be helpful. These should help to point you in the direction that you should be going and even suggest an equation to lead you to the solution.

- If you can, write an equation that will lead to the solution. Key words and phrases in the problem should help lead you to that equation. You'll look at key words and phrases in the next section. Make sure that you don't take extraneous information from the problem and use it in your equation. Sometimes extra information is included in an attempt to confuse you. Be aware of that so you don't fall for it.

- Solve the equation carefully, making sure to check your work as you go. Then see whether the answer you got is the solution to the problem. Sometimes the variable you've found will only be part of what's being asked for.

- Make sure that your answer makes sense in terms of its magnitude and units and the other magnitudes and units in the question. For example, if the problem was about hundreds of miles and your answer is in tens of inches or pounds, that should be a hint that *something is wrong*.

TIP

One of the most common errors made when solving word problems is failing to answer the question that the problem actually asked. Turn the question you were asked into a statement and substitute into it the answer that you found to see whether it works. If the statement makes sense, then you probably solved it correctly.

Techniques for Translating Problems into Equations *(continued)*

Key Words and Phrases

As mentioned earlier, some key words and phrases denote which operation(s) should be performed. I have grouped these phrases below by arithmetic operation. These words or phrases should guide you in determining how your equation(s) should be constructed.

Addition	Subtraction
is increased by . . .	is decreased by . . .
is summed with . . .	is diminished by . . .
is more than . . .	is less than . . .
sums to . . .	remains or remainder . . .
plus . . .	the difference between . . .
is added to . . . *or* in addition to . . .	is reduced by . . .
totals to . . .	is fewer than . . .

Multiplication	Division
the product of . . .	the quotient of . . .
of . . . (like three of 4)	divided by . . .
at . . . (like 5 dozen eggs at $1.29 per dozen)	half . . . (divided by 2)
twice . . . (multiplication by 2)	one-third, one fourth, etc.
multiples of . . .	ratio

TIP

As you work word problems, you'll discover more words that imply or suggest particular arithmetic relations or operations. Remember to be aware of what the question is asking, such as: How fast . . .? How far . . .? How many . . .? How heavy . . .? How much . . .? How old . . .? What percent . . .? Find the ratio. . . . What part of . . .? and so forth.

The problems you are going to study are grouped into types for ease of understanding. Probably the most logical place to start is with number problems, since they relate pure numbers, and there are no other units to worry about. Every solution to a word problem should begin with a "Let statement" that tells what you are letting the variable stand for.

It is always preferable to solve a word problem using an equation with a single variable if it can be done that way. All of the word problems in this book can be solved in terms of one variable.

Number Problems

❶ Two numbers sum to 20. Twice the first number equals three times the second. Find both of the numbers.

Let n = the second number.
Then $20 - n$ = the first number.

It's pretty obvious that what you are looking for here are two numbers. Start the solution to this problem with a **Let statement** (see right). You could let the variable n stand for either the first or the second number. Since the second number is multiplied by a larger amount than the first, I'd recommend letting n stand for the second number. How do you decide what the first number should be? The answer to that is in the first sentence of the problem, "Two numbers sum to 20." If they sum to 20, then 20 minus the second must equal the first.

Since the Let statement was derived from the first sentence, you can't use it to write the equation. The numbers would simply cancel out. Instead, use the problem's second sentence. Read it again, and you'll see that it says what is written in equation form to the right.

$$2(20 - n) = 3n$$

Next, solve the equation for n.

$$2(20 - n) = 3n$$
$$3n = 2(20 - n)$$
$$3n = 40 - 2n$$
$$3n + 2n = 40 - 2n + 2n$$
$$5n = 40$$
$$\frac{5n}{5} = \frac{40}{5}$$
$$n = 8$$

Did you think to go on after finding that $n = 8$? I'll bet you that some people didn't. You have to remember that the problem asked you to find *two* numbers. Try another one.

$$20 - n = 20 - 8$$
$$20 - n = 12$$

❷ When 7 times a number is increased by 9, the result is 37.

Let x = the number.

This is a pretty easy one, because you're only looking for one number, but don't get cocky. Start out with the Let statement.

7 times the number is increased by 9. That translates to:

$$7x + 9$$

When 7 times the number is increased by 9, the result is 37. This last phrase translates to the = sign and the specified quantity.

$$7x + 9 = 37$$

To solve, subtract 9 from each side and then divide by 7.

Does the answer make sense? 7 times 4 is 28 and add 9 to that and get 37. Yep.

$$7x + 9 - 9 = 37 - 9$$
$$7x = 28$$
$$\frac{7x}{7} = \frac{28}{7}$$
$$x = 4$$

Consecutive Integer Problems

There are essentially three types of consecutive integer problems. They are consecutive integers (1, 2, 3, etc.), consecutive odd integers (1, 3, 5, etc.), and consecutive even integers (2, 4, 6, etc.).

❶ The sum of three consecutive integers is 60. Find the numbers.

As always, start with the Let statement.

Let n = the first integer.
Then $n + 1$ = the second.
Then $n + 2$ = the third.

Now go back to the problem to construct the equation.

$$n + n + 1 + n + 2 = 60$$

Next, combine like terms by addition.

$$(n + n + n) + (1 + 2) = 60$$
$$3n + 3 = 60$$
$$3n + 3 - 3 = 60 - 3$$
$$3n = 57$$
$$n = 19$$

Since $n = 19$, $n + 1 = 20$, and $n + 2 = 21$. Do they add up to 60? Check and see.

❷ The sum of three consecutive even integers is 48. What are they? See the right column.

Let n = the first integer.
Then $n + 2$ = the second.
Then $n + 4$ = the third.

Now go back to the problem to construct the equation.

$$n + n + 2 + n + 4 = 48$$

Solving this equation does not take a genius, but be careful.

So, $n = 14$, $n + 2 = 16$, and $n + 4 = 18$.

$$3n + 6 = 48$$
$$3n + 6 - 6 = 48 - 6$$
$$3n = 42$$
$$n = 14$$

....

❸ Twice the smallest of three consecutive odd integers is 1 more than the largest. Find the integers.

The first thing that may surprise you is the Let statement.

It's the same as for three consecutive even integers. Do you get why? Even integers are two apart, and odd integers are also two apart.

Let n = the first integer.
Then $n + 2$ = the second.
Then $n + 4$ = the third.

....

Now, the equation is going to be a bit different.

$$2n = n + 4 + 1$$

The problem reads twice the smallest (that's $2n$) is (that's =) 1 more than the largest (that's $n + 4 + 1$). Next, solve it.

....

So, $n = 5$, $n + 2 = 7$, and $n + 4 = 9$. Is twice the smallest 1 more than the largest? You bet it is.

$$2n - n = n - n + (4 + 1)$$
$$n = 5$$

Age Problems

Age problems are another type of word problem that uses primarily numbers, but which introduces a unit of age, usually years. Your mother may have been 20 when you were born, but when you were 5, she was 5 times as old as you, and five years later she was 3 times as old as you. This changing relationship between the ages of two persons is the basis for some interesting word problems! When considering the meaning of an age problem, remember to keep track of the aging going on between both of the subjects. Remember: A year from now, whatever the relationship between our ages now, I'll be one year older, and so will you.

1 Frank is three times older than Reese. 5 years ago, Frank was 8 times Reese's age. Find their ages now.

A table would come in handy here.

	Ages Now	**Ages 5 Years Ago**
Frank	$3x$	$3x - 5$
Reese	x	$x - 5$

The problem says that 5 years ago Frank's age was 8 times Reese's. Therein lies your equation.

Once you've found Reese's current age, it's not very difficult to find Frank's. Now what did the problem ask for? Checking back, it asks for their ages now.

$$3x - 5 = 8(x - 5)$$
$$3x - 5 = 8x - 40$$
$$8x - 3x - 40 + 40 = 3x - 3x - 5 + 40$$
$$5x = 35$$
$$x = 7$$
$$3x = 21$$

2 Ian is twice as old as Myles. Jake is 4 years younger than Myles. In two years the sum of all of their ages will be 58. How old are they now? Again, a table would be most helpful (see right). Since all the ages seem to be keyed to Myles', let's use an m as the variable.

	Ages Now	**In Two Years**
Myles	m	$m + 2$
Ian	$2m$	$2m + 2$
Jake	$m - 4$	$m - 2$

The equation is to be found in the third sentence of the problem. Read it and then see the equation to the right.

$$m + 2 + 2m + 2 + m - 2 = 58$$

Once again, the solution is nothing much, but be careful not to get mixed up with all the ms and 2s.

That makes Myles 14, Ian 28, and Jake 10.

$$m + 2 + 2m + 2 + m - 2 = 58$$
$$(m + 2m + m) + (2 + 2 - 2) = 58$$
$$4m + 2 = 58$$
$$4m + 2 - 2 = 58 - 2$$
$$4m = 56$$
$$m = 14$$

Types of Problems
(continued)

Distance and Motion Problems

This is the first type of problem in which drawing a diagram might do you some good. This is also the first type of problem that can be referred to as **formulaic**. That is, there is a standard formula to be followed for most situations. Either straight or in some variation, that formula is shown at right:

distance = rate × time, or $d = rt$

❶ An automobile travels 42 km at a speed of 84 km/hr (that's kilometers per hour). How long did it take?

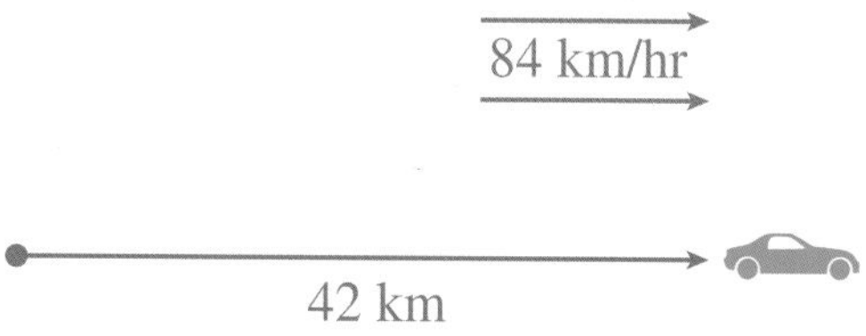

The problem tells you the distance and the rate (speed); what you need to find is the time.

84 km/hr

42 km

Next, substitute what you know into the equation.

It took the automobile $\frac{1}{2}$ hour. Does that make sense? The car was traveling at twice the km per hour as the distance it traversed, so it does make sense.

$d = rt$

$42 = 84t$

$\dfrac{84t}{84} = \dfrac{42}{84}$

$t = \dfrac{1}{2}$

❷ Rocio left New York City at noon and drove west on I-80 at 60 miles per hour. Kira left New York City an hour later and drove west on I-80 at 70 mph. Both continue to drive without stopping and at the same speeds. At what time will Kira catch up to Rocio, and how far will each have driven?

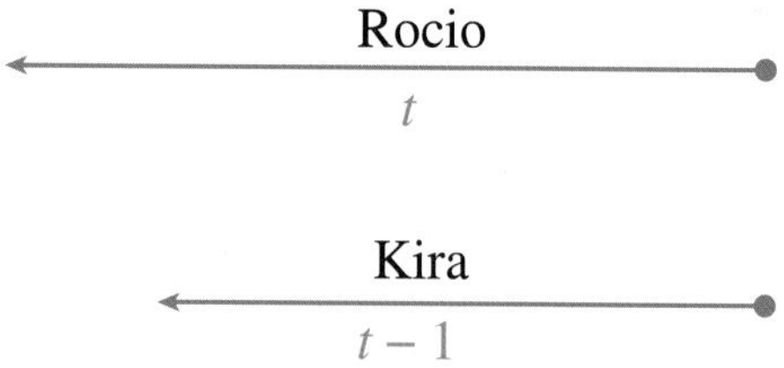

Rocio

t

Kira

$t - 1$

A table will definitely help.

	Rate ...*r*	Time ...*t*	Distance ...*d*
Rocio	60 mph	*t* hrs	60*t* mi.
Kira	70 mph	*t* – 1 hrs	70(*t* – 1) mi.

The I-80 information is gratuitous and was thrown in to confuse matters. The times that the two of them drive are different. So are their speeds, but when Kira catches up to Rocio, both will have driven the same distance, and that's how you get the equation to the right:

$$60t = 70(t - 1)$$
$$60t = 70t - 70$$
$$60t - 70t = 70t - 70t - 70$$
$$-10t = -70$$
$$t = 7$$

Notice that you started by equating distances and ended up finding the time. There it is. *t* = 7 hrs. Does that answer the question? Well, the question was at what time did Kira catch up. A look at the chart will show you that *t* represents Rocio's time, and she left at noon, so noon + 7 hours makes it 7 P.M. That answers part of the question. The second part is how far each has driven. The easier equation to solve will be for Rocio.

$$d = rt$$
$$d = 60 \cdot 7$$
$$d = 420 \text{ miles}$$

And now you've answered the questions that were asked.

Coin Problems

Coin problems involve nickels, dimes, quarters, and occasionally half dollars and pennies. They all tell you something about the total number of coins, the relationships in quantities of each type of coin, and the total value of the coins. It is then your job to find out how many coins of each type are involved.

1. Tania has a number of coins totaling $3.00. She has twice as many dimes as quarters and 5 more nickels than dimes. How many coins of each type does she have?

Types of Problems
(continued)

There are three procedures to follow when working with coin problems.

Work in terms of pennies, so a nickel is worth 5 cents rather than $.05, a dime is worth 10 cents, and $3.00 is 300 cents.

Always pick the smallest quantity to be the variable. In this case, that would be the quarters.

Always make good use of a table. You might have noticed by now that a table is nothing more than a fancy form of a Let statement:

Coins	Number	Value of Each (in cents)	Total Value
nickels	$2x + 5$	5	$10x + 25$
dimes	$2x$	10	$20x$
quarters	x	25	$25x$

Now you're ready to put together an equation based on the total value of Tania's coins.

$$25x + 20x + 10x + 25 = 300$$

Finally, combine terms and solve for the number of quarters.

$$25x + 20x + 10x + 25 = 300$$
$$55x + 25 = 300$$
$$55x + 25 - 25 = 300 - 25$$
$$55x = 275$$
$$x = 5$$

Now that you have the number of quarters, 5, you know there are twice as many dimes, or 10, and 5 more nickels than that, or 15 nickels. You figure out whether that makes $3.00.

② Sebastian has 102 coins in quarters and dimes that total $21.00. How many of each coin does he have?

Remember the three procedures you followed last time, as you construct the chart. You can let either coin be the variable. Use d for dimes. If d stands for dimes, how many quarters are there?

Coins	Number	Value of Each (in cents)	Total Value
dimes	d	10	$10d$
quarters	$102 - d$	25	$2550 - 25d$

Now you're ready to put together an equation based on the total value of Sebastian's coins.

$$10d + 2550 - 25d = 2100$$

Next, collect like terms and combine them to solve for d.

So there are 30 dimes. How many quarters are there? Well, since $102 - 30 = 72$, that was an easy one!

$$10d - 25d + 2550 - 2550 = 2100 - 2550$$
$$-15d = -450$$
$$\frac{-15d}{-15} = \frac{-450}{-15}$$
$$d = 30$$

Mixture Problems

Mixture problems deal with having two or more kinds of things mixed together, be it solutions of different concentrations, nuts in a can that sell for two (or more) different prices, more expensive and less expensive coffees, etc.

There are generally two matters of consequence in dealing with mixtures. They are the amount of each and the value or concentration of each. After both of these are considered, the result is the mixture of those items combined.

$$a\% \text{ solution} + b\% \text{ solution} + c\% \text{ solution}$$

x amount y amount $(x + y)$ amount

Here is the first of two examples.

1 Solution 1 is 75% sodium hydroxide (formula NaOH), while Solution 2 is 50% sodium hydroxide. How many fl. oz. (fluid ounces) of each should be used in order to end up with 50 fl. oz. of a 60% sodium hydroxide solution?

	% NaOH	fl. oz.	Concentration of NaOH
Solution 1	75%	x	$.75x$
Solution 2	50%	$50 - x$	$.50(50 - x)$
Mixture	60%	50	$.60(50)$

Two things should be pointed out here. First of all, $.5$ and $.6$ could have been used in place of $.50$ and $.60$, respectively, in the last column. The superfluous zeroes were kept to avoid confusing you. The second thing to note is the last column contains the equation you need to solve. Just rewrite that addition in horizontal form.

$$.75x + .50(50 - x) = .60(50)$$

Now, proceed to combine like terms and to isolate the variable.

That's 20 fl. oz. of Solution 1 and 30 fl. oz. of Solution 2.

$$.75x + .50(50 - x) = .60(50)$$
$$.75x + 25 - .5x = 30$$
$$.25x + 25 - 25 = 30 - 25$$
$$100(.25x = 5)$$
$$25x = 500$$
$$x = 20$$
$$50 - x = 30$$

Now check out this one.

❷ 12 pounds of pecans that cost $4.75 per pound are to be mixed with 16 lbs of raisins that cost $1.25/lb (/lb = per pound). What should be charged per pound for the mixture?

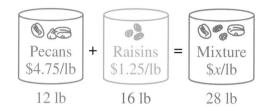

	Cost/lb	Pounds	Total Cost
Pecans	$4.75	12	12 (4.75)
Raisins	$1.25	16	16(1.25)
Mixture	x	28	28x

Again, the equation is in the rightmost column of the table.

$$28x = 16(1.25) + 12(4.75)$$

Now combine like terms and solve.

This tells you that the mixture should go for $2.75 per pound.

$$28x = 16(1.25) + 12(4.75)$$
$$28x = 20 + 57$$
$$\frac{28x}{28} = \frac{77}{28}$$
$$x = 2.75$$

Work Problems

Work problems consider the situations in which two or more people work together to complete a single job. The general pattern is, for example, one person can do a job working alone in three hours, which means that person could do one third of the job in an hour. A second person could do the job working alone in 2 hours, which means that person could do one half of the job in an hour. A third person working alone could do the job in 4 hours, which means that person could do one fourth of the job in an hour. How long would it take to do the job if they all worked together?

The solution to this type of problem takes the following form:

$$\frac{1}{1 \text{ person's time}} + \frac{1}{2\text{nd person's time}} + \frac{1}{3\text{rd person's time}} + \text{etc.} = \frac{1}{x}$$

Use the figures from the first paragraph and substitute them into this formula—omitting the "etc." this time.

$$\frac{1}{1 \text{ person's time}} + \frac{1}{2\text{nd person's time}} + \frac{1}{3\text{rd person's time}} = \frac{1}{x}$$

$$\frac{1}{3} + \frac{1}{2} + \frac{1}{4} = \frac{1}{x}$$

In order to combine those fractions, you need a common denominator, the least of which happens to be 12. With that in mind, convert all three fractions to 12ths.

$$\frac{1}{3} + \frac{1}{2} + \frac{1}{4} = \frac{1}{x}$$

$$\frac{4}{12} + \frac{6}{12} + \frac{3}{12} = \frac{1}{x}$$

$$\frac{13}{12} = \frac{1}{x}$$

Next, cross-multiply and solve for x.

That means that the three of them working together could finish the job in that fraction of an hour.

$$\frac{13}{12} = \frac{1}{x}$$
$$13x = 12$$
$$\frac{13x}{13} = \frac{12}{13}$$
$$x = \frac{12}{13}$$

Suppose Amy, working alone, could paint her entire house in 3 days, Beth (even though she's not feeling well) could paint it alone in 6 days, and Jo alone would take 9 days (admittedly, they're little women). How long would it take them to paint the house if they all worked together? Based on what you just learned:

$$\frac{1}{3} + \frac{1}{6} + \frac{1}{9} = \frac{1}{x}$$

The least common denominator for the three fractions is 18; therefore, convert and combine.

$$\frac{6}{18} + \frac{3}{18} + \frac{2}{18} = \frac{1}{x}$$
$$\frac{11}{18} = \frac{1}{x}$$

Finally, cross-multiply and solve for x.

$$\frac{11}{18} = \frac{1}{x}$$
$$11x = 18$$
$$\frac{11x}{11} = \frac{18}{11}$$
$$x = \frac{18}{11} \text{ or } 1\frac{7}{11} \text{ days}$$

You'll get the opportunity to practice all of these types of word problems in the practice section that's coming right up.

Practice Questions

Solve each problem to the best of your ability. If applicable, express the answer in simplest possible terms.

1 When 8 times a number is decreased by 7, the result is 49. What is the number?

2 Two numbers sum to 18. Twice the first number equals four times the second. Find both of the numbers.

3 One number exceeds another by 7. If the sum of twice the numbers is 46, find the numbers.

4 The sum of three consecutive integers is 75. Find the numbers.

5 The sum of three consecutive even integers is 102. What are they?

6 The sum of the smallest and largest of three consecutive odd integers is thirteen greater than the middle one. Find the integers.

7 The second integer is triple the smallest, and largest is triple that. They sum to 208. Find the integers.

8 Carmen is four times older than Mariel. 6 years ago, Carmen was 7 times Mariel's age. Find their ages now.

9 Melissa is three times as old as Frank. Gina is 6 years younger than Frank. In two years the sum of all of their ages will be 50. How old are they now?

10 Alex's grandfather is 30 times as old as Alex is. In 16 years he'll be 4 more than 4 times Alex's age. How old are Alex and his grandfather now?

11 Anna is twice Marcello's age and 6 years younger than Ali. In 3 years their ages will sum to 50. How old is each of them now?

12 An automobile travels 126 km at a speed of 84 km/hr. How long did it take?

13 Andy left Chicago at 9 A.M. and drove south on a four-lane highway at a steady 40 miles per hour. Fil left Chicago a half hour later and drove south on the same road at 50 mph. Both continue to drive without stopping at the same speeds. At what time will Fil catch up to Andy, and how far will each have driven?

⑭ A bus left the terminal at 11 A.M. Another bus traveling the same route left the same terminal at 12:30 P.M. and caught up to the first bus at 6:00 P.M. If the first bus was traveling 12 mph slower than the second one, how fast was each bus going?

⑮ A truck leaves Chicago and travels north at 60 mph. An hour earlier a truck left Chicago traveling south. When will they be 675 miles apart?

⑯ Melissa has a number of coins totaling $4.00. She has twice as many dimes as nickels and 4 more nickels than quarters. How many coins of each type does she have?

⑰ Francesco has equal numbers of quarters and dimes, and half as many nickels. Altogether he has a total of $7.50. How many of each coin does he have?

⑱ Solution A is 80% sulfuric acid (formula H_2SO_4), while Solution B is 40% sulfuric acid. How many ml. of each should be used in order to end up with 200 ml of a 50% sulfuric acid solution?

⑲ $2.10 per pound coffee is mixed with $1.70 per pound coffee to make 20 pounds of a blend that will sell for $1.80 per pound. How many pounds of each coffee will go into the blend?

⑳ How many fluid ounces of chocolate syrup must be added to a pint of milk to make a solution that's 20% chocolate syrup by volume?

㉑ Greg can varnish his living room floor in about four hours. His friend, Suzanne, can do it in three. How long would it take to varnish Greg's living room floor if Greg and Suzanne worked together?

㉒ David can peel 5 pounds of potatoes in 15 minutes. Karen can peel 5 pounds of potatoes in 10 minutes. How long would it take them to peel 5 pounds of potatoes if they both worked at it together?

㉓. Working alone, Jason can seal a driveway in 2 hours; Ian can seal a driveway in 4 hours; Dylan can seal a driveway in 5 hours. How long would it take them to seal the driveway if all three of them worked together?

① When 8 times a number is decreased by 7, the result is 49. What is the number?

Let x = the number.

8 times the number, decreased by 7, translates to:

$$8x - 7 = 49$$

Now collect and combine like terms to solve that equation:

$$8x - 7 = 49$$
$$8x - 7 + 7 = 49 + 7$$
$$8x = 56$$
$$\frac{8x}{8} = \frac{56}{8}$$
$$x = 7$$

Does the answer make sense? 8 times 7 is 56; subtract 7 from that and get 49. That works.

② Two numbers sum to 18. Twice the first number equals four times the second. Find both of the numbers.

First comes the Let statement.

Let n = the second number.

Then $18 - n$ = the first number.

Use the problem's second sentence to derive the equation:

$$2(18 - n) = 4n$$

Next, solve the equation for n.

$$2(18 - n) = 4n$$
$$4n = 2(18 - n)$$
$$4n = 36 - 2n$$
$$6n = 36$$
$$\frac{6n}{6} = \frac{36}{6}$$
$$n = 6$$
$$18 - n = 18 - 6$$
$$18 - n = 12$$

3 One number exceeds another by 7. If the sum of twice the numbers is 46, find the numbers.

Let x = the first integer.

Then $x + 7$ = the second.

Now, from the second sentence, form the equation:

$$2x + 2(x + 7) = 46$$

Now clear parentheses, collect like terms, combine, and solve:

$$2x + 2(x + 7) = 46$$
$$4x + 14 - 14 = 46 - 14$$
$$4x = 32$$
$$\frac{4x}{4} = \frac{32}{4}$$
$$x = 8$$
$$x + 7 = 15$$

④ The sum of three consecutive integers is 75. Find the numbers.

Let n = the first integer.

Then $n + 1$ = the second.

Then $n + 2$ = the third.

Now go back to the problem to construct the equation:

$$n + n + 1 + n + 2 = 75$$

Next, collect and combine like terms:

$$(n + n + n) + (1 + 2) = 75$$
$$3n + 3 = 75$$
$$3n + 3 - 3 = 75 - 3$$
$$3n = 72$$
$$n = 24$$
$$n + 1 = 25$$
$$n + 2 = 26$$

⑤ The sum of three consecutive even integers is 102. What are they?

Let n = the first integer.

Then $n + 2$ = the second.

Then $n + 4$ = the third.

Then go back to the problem to construct the equation:

$$n + n + 2 + n + 4 = 102$$

Now solve carefully:

$$(n + n + n) + (4 + 2) = 102$$
$$3n + 6 = 102$$
$$3n + 6 - 6 = 102 - 6$$
$$3n = 96$$
$$n = 32$$
$$n + 2 = 34$$
$$n + 4 = 36$$

6 The sum of the smallest and largest of three consecutive odd integers is thirteen greater than the middle one. Find the integers.

Let n = the first integer.

Then $n + 2$ = the second.

Then $n + 4$ = the third.

Remember, even integers are two apart, and odd integers are also two apart. Next, the equation: The sum of the smallest and largest (that's $n + n + 4$) is 13 greater than the middle one (= $n + 2 + 13$):

$$n + n + 4 = n + 2 + 13$$

Next, solve it.

$$2n + 4 = n + 15$$
$$2n - n + 4 - 4 = n - n + 15 - 4$$
$$n = 11$$
$$n + 2 = 13$$
$$n + 4 = 15$$

7 The second integer is triple the smallest, and the largest is triple that. They sum to 208. Find the integers.

Let x = the first integer.

Then $3x$ = the second.

Then $9x$ = the third.

The equation is quite straightforward:

$$x + 3x + 9x = 208$$
$$13x = 208$$
$$x = 16$$
$$3x = 48$$
$$9x = 144$$

8 Carmen is four times older than Mariel. 6 years ago, Carmen was 7 times Mariel's age. Find their ages now.

A table is needed here.

Girls	Ages Now	Ages 6 Years Ago
Carmen	$4m$	$4m - 6$
Mariel	m	$m - 6$

The problem says that 6 years ago Carmen's age was 7 times Mariel's. Therein lies your equation:

$$4m - 6 = 7(m - 6)$$
$$4m - 6 = 7m - 42$$
$$7m - 42 = 4m - 6$$
$$7m - 4m - 42 + 42 = 4m - 4m - 6 + 42$$
$$3m = 36$$
$$m = 12$$
$$4m = 48$$

9 Melissa is three times as old as Frank. Gina is 6 years younger than Frank. In two years the sum of all of their ages will be 50. How old are they now?

Since all the ages are keyed to Frank's age, use an *f* as the variable.

Children	Ages Now	Ages in Two Years
Frank	f	$f + 2$
Melissa	$3f$	$3f + 2$
Gina	$f - 6$	$f - 4$

The equation is to be found in the third sentence of the problem:

$$f + 2 + 3f + 2 + f - 4 = 50$$

Now collect and combine terms.

$$f + 2 + 3f + 2 + f - 4 = 50$$
$$\left(f + 3f + f\right) + \left(2 + 2 - 4\right) = 50$$
$$5f = 50$$
$$\frac{5f}{5} = \frac{50}{5}$$
$$f = 10$$
$$3f = 30$$
$$f - 6 = 4$$

That makes Frank 10, Melissa 30, and Gina 4.

10 Alex's grandfather is 30 times as old as Alex is. In 16 years he'll be 4 more than 4 times Alex's age. How old are Alex and his grandfather now?

Since all the ages are keyed to Alex's age, use an a as the variable.

	Ages Now	Ages in 16 Years
Alex	a	$a + 16$
Alex's grandfather	$30a$	$30a + 16$

The equation is to be found in the second sentence of the problem:

$$30a + 16 = 4(a + 16) + 4$$
$$30a + 16 = 4a + 64 + 4$$
$$30a - 4a + 16 - 16 = 4a - 4a + 68 - 16$$
$$26a = 52$$
$$a = 2$$
$$30a = 60$$

Alex is 2; his grandfather is 60.

11 Anna is twice Marcello's age and 6 years younger than Ali. In 3 years their ages will sum to 50. How old is each of them now?

	Ages Now	Ages in Three Years
Marcello	x	$x + 3$
Ali	$2x + 6$	$2x + 9$
Anna	$2x$	$2x + 3$

Note that since Anna is 6 years younger than Ali, Ali is 6 years older than she. The equation is to be found in the second sentence of the problem:

$$x + 3 + 2x + 9 + 2x + 3 = 50$$
$$(x + 2x + 2x) + (3 + 9 + 3) = 50$$
$$5x + 15 - 15 = 50 - 15$$
$$5x = 35$$
$$x = 7$$
$$2x = 14$$
$$2x + 6 = 20$$

That makes Marcello 7, Anna 14, and Ali 20.

12 An automobile travels 126 km at a speed of 84 km/hr. How long did it take? The problem tells you the distance and the rate; what you need to find is the time.

Next, substitute what you know into the distance equation:

$$d = rt$$
$$126 = 84t$$
$$\frac{84t}{84} = \frac{\cancel{126}^{3}}{\cancel{84}_{2}}$$
$$t = \frac{3}{2}$$

That is $1\frac{1}{2}$ hours.

⑬ Andy left Chicago at 9 A.M. and drove south on a four-lane highway at a steady 40 miles per hour. Fil left Chicago a half hour later and drove south on them same road at 50 mph. Both continue to drive without stopping at the same speeds. At what time will Fil catch up to Andy, and how far will each have driven?

A table will definitely help.

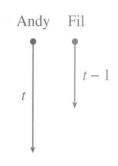

Andy   Fil

t

$t - 1$

	Rate ...r	Time ...t	Distance ...d
Andy	40 mph	t hrs	$40t$ mi.
Fil	50 mph	$t - \frac{1}{2}$ hrs	$50(t - \frac{1}{2})$ mi.

The essential part to remember is that when Fil catches Andy, both will have driven the same distance; that's how you get your equation:

$$40t = 50\left(t - \frac{1}{2}\right)$$
$$40t = 50t - 25$$
$$40t - 50t = 50t - 50t - 25$$
$$-10t = -25$$
$$t = 2.5$$

2.5 hours added to Andy's starting time of 9 A.M. yields a catching up time of 11:30 A.M.

You still need to find the distance driven:

$$d = rt$$
$$d = 40 \cdot 2.5$$
$$d = 100 \text{ miles}$$

And now you've answered both questions that were asked.

⑭ A bus left the terminal at 11 A.M. Another bus traveling the same route left the same terminal at 12:30 P.M. and caught up to the first bus at 6:00 P.M. If the first bus was traveling 12 mph slower than the second one, how fast was each bus going?

	Rate ...r	Time ...t	Distance ...d
Bus 1	r	7 hrs.	$7r$ mi.
Bus 2	$r + 12$	5.5 hrs.	$5.50(r + 12)$ mi.

Form the equation by setting the two distances equal to each other; after all, both buses traveled the same distance.

$$7r = 5.5(r + 12)$$
$$7r = 5.5r + 66$$
$$7r - 5.5r = 5.5r - 5.5r + 66$$
$$1.5r = 66$$
$$r = 44$$
$$r + 12 = 56$$

They were traveling at 44 and 56 mph, respectively.

⑮ A truck leaves Chicago and travels north at 60 mph. An hour earlier a truck left Chicago traveling south. When will they be 675 miles apart?

	Rate ...r	Time ...t	Distance ...d
Truck 1 (earlier)	45	t hrs.	$45t$ mi.
Truck 2 (later)	60	$t - 1$ hrs.	$60(t - 1)$ mi.

You're looking for a time when the total distance between them will be 675 miles, so:

$$45t + 60(t - 1) = 675$$
$$45t + 60t - 60 = 675$$
$$105t - 60 + 60 = 675 + 60$$
$$105t = 735$$
$$t = 7$$

That's 7 hours, even.

16 Melissa has a number of coins totaling $4.00. She has twice as many dimes as nickels and 4 more nickels than quarters. How many coins of each type does she have?

Coins	Number	Value of Each (in cents)	Total Value
nickels	$x + 4$	5	$5x + 20$
dimes	$2(x + 4)$	10	$20(x + 4)$
quarters	x	25	$25x$

Now you're ready to put together an equation based on the total value of Melissa's coins:

$$5x + 20 + 20(x + 4) + 25x = 400$$

Remember, everything's in terms of pennies.

$$5x + 20 + 20(x + 4) + 25x = 400$$
$$5x + 20 + 20x + 80 + 25x = 400$$
$$50x + 100 = 400$$
$$50x + 100 - 100 = 400 - 100$$
$$50x = 300$$
$$x = 6$$

Now that you have the number of quarters, 6, you know there are 4 more nickels, or 10, and twice as many dimes as nickels, 20.

⑰ Francesco has equal numbers of quarters and dimes, and half as many nickels. Altogether, he has a total of $7.50. How many of each coin does he have?

Coins	Number	Value of Each (in cents)	Total Value
nickels	x	5	$5x$
dimes	$2x$	10	$20x$
quarters	$2x$	25	$50x$

Now you're ready to put together an equation based on the total value of Francesco's coins:

$$5x + 20x + 50x = 750$$

Next, combine terms and solve for x:

$$5x + 20x + 50x = 750$$
$$75x = 750$$
$$\frac{75x}{75} = \frac{750}{75}$$
$$x = 10$$

So there are 10 nickels. How many quarters and dimes are there? Well, since he has half as many nickels as quarters and dimes, he must have twice as many of those. That's 20 of each.

⑱ Solution A is 80% sulfuric acid (formula H_2SO_4), while Solution B is 40% sulfuric acid. How many ml. of each should be used in order to end up with 200 ml of a 50% sulfuric acid solution?

	% H_2SO_4	ml	Concentration of H_2SO_4
Solution A	80%	x	$.80x$
Solution B	40%	$200 - x$	$.40(200 - x)$
Mixture	50%	200	$.50(200)$

Remember that the rightmost column contains the equation:

$$.80x + .40(200 - x) = .50(200)$$
$$10[.80x + .40(200 - x) = .50(200)]$$
$$8x + 4(200 - x) = 5(200)$$

By multiplying each term by 10, you get rid of the necessity to work with decimals. Next, proceed to combine like terms and to isolate the variable:

$$8x + 4(200 - x) = 5(200)$$
$$8x + 800 - 4x = 1000$$
$$4x + 800 - 800 = 1000 - 800$$
$$4x = 200$$
$$\frac{4x}{4} = \frac{200}{4}$$
$$x = 50$$

That's 50 ml of Solution A and 150 ml of Solution B.

⓳ $2.10 per pound coffee is mixed with $1.70 per pound coffee to make 20 pounds of a blend that will sell for $1.80 per pound. How many pounds of each coffee will go into the blend?

	Cost/lb	Pounds	Total Cost
$2.10/lb coffee	$2.10	x	$2.1x$
$1.70/lb coffee	$1.70	$20 - x$	$1.7(20 - x)$
Mixture	$1.80	20	$1.8(20)$

Again, the equation is in the rightmost column. Note that you've elected to drop the superfluous zero that would otherwise follow each decimal point:

$$2.1x + 1.7(20 - x) = 1.8(20)$$
$$10[2.1x + 1.7(20 - x) = 1.8(20)]$$
$$21x + 17(20 - x) = 18(20)$$

Once more, multiply through by 10 to get rid of the decimals altogether. Now combine like terms and solve.

$$21x + 17(20 - x) = 18(20)$$
$$21x + 340 - 17x = 360$$
$$4x + 340 - 340 = 360 - 340$$
$$4x = 20$$
$$\frac{4x}{4} = \frac{20}{4}$$
$$x = 5$$

That means you'll use 5 pounds of the more expensive coffee and 15 pounds of the less expensive one.

20 How many fluid ounces of chocolate syrup must be added to a pint of milk to make a solution that's 20% chocolate syrup by volume?

	% Syrup	Fl. Oz.	Concentration of Syrup
Syrup	100%	x	$1.0(x)$
Milk	0%	16	$0(16)$
Mixture	20%	$16 + x$	$.2(16 + x)$

Note that you need to change 1 pint to 16 fl. oz. Remember also that the rightmost column contains the equation:

$$1.0x + 0(16) = .2(16 + x)$$
$$10[1.0x + 0(16) = .2(16 + x)]$$
$$10x + 0(16) = 2(16 + x)$$

By multiplying each term by 10, you get rid of the necessity to work with decimals. Next, proceed to combine like terms and to isolate the variable:

$$10x + 0 = 32 + 2x$$
$$10x - 2x = 32 + 2x - 2x$$
$$8x = 32$$
$$\frac{8x}{8} = \frac{32}{8}$$
$$x = 4$$

That means four fluid ounces of chocolate syrup are required.

21 Greg can varnish the living room floor in about four hours. His sister, Suzanne, can do it in three. How long would it take to varnish Greg's living room floor if Greg and Suzanne worked together?

First, write the formula:

$$\frac{1}{4} + \frac{1}{3} = \frac{1}{x}$$

For thirds and fourths, the LCD is twelfths, so convert both fractions and add.

$$\frac{3}{12} + \frac{4}{12} = \frac{1}{x}$$
$$\frac{7}{12} = \frac{1}{x}$$
$$7x = 12$$
$$x = \frac{12}{7}, \text{ or } 1\frac{5}{7} \text{ hours}$$

㉒ David can peel 5 pounds of potatoes in 15 minutes. Karen can peel 5 pounds of potatoes in 10 minutes. How long would it take them to peel 5 pounds of potatoes if they both worked at it together?

First construct the formula and then find the common denominator, combine, cross-multiply, and solve:

$$\frac{1}{15} + \frac{1}{10} = \frac{1}{x}$$

$$\frac{2}{30} + \frac{3}{30} = \frac{1}{x}$$

$$\frac{5}{30} = \frac{1}{x}$$

$$5x = 30$$

$$x = 6 \text{ minutes}$$

㉓ Working alone, Jason can seal a driveway in 2 hours; Ian can seal a driveway in 4 hours; Dylan can seal a driveway in 5 hours. How long would it take them to seal the driveway if all three of them worked together?

First construct the formula; then you'll find the common denominator, combine, cross-multiply, and solve:

$$\frac{1}{2} + \frac{1}{4} + \frac{1}{5} = \frac{1}{x}$$

$$\frac{10}{20} + \frac{5}{20} + \frac{4}{20} = \frac{1}{x}$$

$$\frac{19}{20} = \frac{1}{x}$$

$$19x = 20$$

$$x = \frac{20}{19}; \text{ or } 1\frac{1}{19} \text{ hours}$$

Glossary of Terms

absolute value   The distance of a number from 0, denoted by $\|$ brackets: $|-3| = |3| = 3$

axes   The two lines, one horizontal and the other vertical, that cross at the origin

axiom   An obvious statement or rule that is accepted without need for proof

binomial   An expression containing two terms, separated by a plus or a minus sign

Cartesian coordinates   The name given the graphing plane denoted by the x- and y- axes, so-named in honor of French philosopher/mathematician, Rene Descartes

coefficient   One of two or more numbers multiplied together in a single term

collect like terms   In an equation, the process of getting all terms containing variables on one side of the equal sign and all constants on the other

common fraction   A fraction with a numerator and a denominator

constant   A number that always has the same value, such as 2, 3, etc.

decimal   A short way to refer to a decimal fraction

decimal fraction   A fraction expressed as a power of 10 by use of a decimal point

decimal point   A dot used to separate whole numbers from fractions in decimal notation

denominator   The bottom part of a common fraction

difference of two squares   The binomial containing two perfect squares separated by a minus sign, which factors to be the sum times the difference of the two square roots, for example, $a^2 - b^2 = (a + b)(a - b)$

direct variation   The increase of one variable as another increases

distributive property   The property relating multiplication and addition through the formula $a(b + c) = ab + ac$; the a is said to have been distributed over the b and c

dividend   In a division, the number being divided into

divisor   In a division, the number being divided by

domain All of the possible input values of a relation

element Synonymous with a member of a set

equal Of the same exact quantitative value

equal sets Sets containing exactly the same elements

equation A relationship between two expressions of equal value

equilibrium In balance; in physics, said of forces acting on any object

equivalent sets Sets containing exactly the same *number* of elements

factor (*n.*) One of two or more numbers multiplied together; (*vt.*) to separate by division

fraction Any rational number; part of a whole; a ratio

function A relation between a dependent and an independent variable such that for each of the latter there is only one of the former

inequality A relationship between two expressions of usually unequal value, denoted by the signs $<$, $\leq$, $>$, and $\geq$

intercept The point(s) at which a graph crosses the y-axis or the x-axis

intersection of sets The collection of elements common to two or more sets

inverse variation The increase of one variable as another decreases

linear equation Contains no terms raised to a power above 1; graphs as a straight line

magnitude The quantitative size of something

member Synonymous with element of a set

monomial An algebraic expression that consists of only a single term

numerator The top part of a common fraction

origin The point with coordinates $(0, 0)$, at which the x-axis and y-axis cross

percent A fraction based on 100 being one whole; represented with the % sign

plane A flat surface that continues forever in two directions; see *Cartesian coordinates*

polynomial An expression containing many terms; a binomial is the smallest example of a polynomial

proportion An equating of one ratio to another

quadratic equation An equation containing a term raised to the second power

range All of the possible output values of a relation

ratio A comparison of one thing to a second in which order matters

reciprocal Of a number, what that number must be multiplied by to get a product of 1

reflexivity The notion that anything is equal to itself

relation A connection between two variables usually shown as an equation or graph

simultaneous equations Two or more equations in two or more variables that must be solved together; also known as systems of equations

slope The slant of a graphed line or at any point on a curve; rise over run

symmetry A property illustrated by if $a = b$ then $b = a$

system of equations See *simultaneous equations*

term Any collection of constants and variables not separated by a + or − sign

transitivity A property illustrated by if $a = b$ and $b = c$, then $a = c$

union of sets A joining of two or more sets that contains all the elements of each of them

variable A letter that is used to represent a quantity; it has the same value within a single mathematical situation, but may vary in value from one situation to another

variation The relationship between one set of variables and another; see also *direct* and *inverse variation*

Venn diagram A circle diagram in which each circle corresponds to the boundary of a set

Index

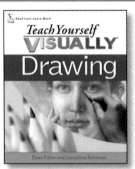

designed for visual learners like you!

Make beautiful music

Make Rover behave

978-0-470-04850-4

978-0-7645-9642-1

978-0-471-74990-5

978-0-471-74989-9

Make improvements in your game

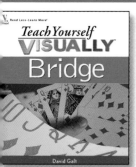

978-0-470-11424-7

978-0-470-04983-9

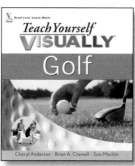

978-0-470-09844-8

978-0-471-79906-1

Make the most of technology

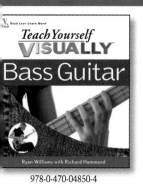

978-0-470-16878-3

978-0-470-04595-4

978-0-470-11452-0

978-0-470-04573-2